# The Living Dock

Other Books by Jack Rudloe

**The Sea Brings Forth**
**The Erotic Ocean**
**Time of the Turtle**
**The Wilderness Coast**

# The Living Dock

## Jack Rudloe

Illustrations from the works of

## Walter Inglis Anderson

Fulcrum, Inc.
Golden, Colorado

Library of Congress Cataloging-in-Publication Data

Rudloe, Jack.
    The Living Dock

    Includes Index.
    1. Marine fauna—Florida—Collection and preservation.
    I. Anderson, Walter Inglis, 1903-1965.
    II. Title

    QL127.R79    1988    591.02'075    88-19963
    ISBN 1-55591-036-X

Some of the characters in the book are fictitious, others are
not. The setting is for the most part real.

Fulcrum, Inc.
Golden, Colorado

To
Nixon Griffis, Colin Phipps,
Nat Pulsifer, Dick Wakefield,
Frank Lehn, and the memory
of Melanie Phipps—
who all know why, or should.

## Acknowledgments

Over the years a number of people have given me inspiration, encouragement and assistance in the writing of *The Living Dock*. Anne, my wife, has spent long, tedious hours taking photographs and reading over the manuscript. Her major professor, Dr. William Herrnkind, patiently listened to my ideas in their early conceptual stages and later plowed through the roughest of drafts. And my mother, Florence Rudloe, was a great help in proofreading and making suggestions.

There isn't enough space available to list all the scientists and specialists who have assisted me with identifications of creatures that were caught off the dock. However, I am grateful to the late Dr. Robert J. Menzies of Florida State University for his hours of discussion with me on fouling communities. Dr. Willard D. Hartman of Yale's Peabody Museum of Natural History has been a source of endless help to me over the years, particularly with sponge classifications, and the same is true of Dr. Charles E. Cutress of the Institute of Marine Science at Mayaguez, Puerto Rico, who identified our anemones.

In an endeavor like this there are a lot of people who have lent their assistance in a variety of ways. I must thank Dr. Philip Greear of Shorter College for sharing his ideas on prehistoric man and his dependence upon the sea. James Dickey has encouraged me to go on writing when the future looked bleak. And my secretary, Mary Ellen Chastain, deserves special thanks because she has been the backbone of Gulf Specimen Company, enabling me to spend long hours at the typewriter. Doug Gleeson, my assistant, has been a rock of dependability, and Leon Crum, a source of both jollity and help.

But I must give full credit to the late Peter V. Ritner, my former editor at World, who turned this book from another spin-off idea into a reality. "If Thoreau can write a book about a pond, you can write one about a dock." The idea seemed totally preposterous at the time; then it grew on me. And so here it is.

## A Special Note of Thanks, 1988

Society and the art world owe a great debt to Mary Anderson Pickard for her lifetime of curating and preserving her father's works. For without her painstaking hours of sorting, cataloging and protecting his oeuvre—a costly and difficult process—most of it would have long ago deteriorated in the sweltering, humid Mississippi heat, and long ago rotted into detritus and joined the web of life. It would have been lost to us all.

I shall always be grateful to her and the rest of the Anderson family for their encouragement, support and hospitality in Ocean Springs; for allowing me to stay in Walter Anderson's cottage, fostering my creativity, giving me inspiration—and, most important, for just being friends.

## Author's Note, 1988

There have been too many changes since I first wrote the *The Living Dock at Panacea* in 1977. With fifty thousand people pouring into Florida each year, and no sign that it is going to stop, much of the wildness of the Florida coast is disappearing. To a naturalist it's a gloomy prospect, as each passing year brings fewer trees, more shopping centers and subdivisions.

The Panhandle is all that is left of the old Florida, although the paper companies have taken their toll on the forests there. St. Joe Bay, with all its biodiversity, clear water and grass beds is now slated for condominiums, marinas and sprawling development. Sewer and water lines are being run. And it seems every morning I wake up to the sounds of bulldozers in Panacea turning wetlands into drylands.

Still, we have won a few conservation battles, and there are still trees and marshes around Dickerson Bay. They exist because land values haven't climbed high enough. The marina wasn't built only because of a hard-won battle. The State of Florida, the Nature Conservancy and the Trust for Public Lands have been desperately trying to purchase blocks of wildlands, but as the population swells and property values escalate, it becomes a cliff-hanging race. Will any wilderness survive, or will the developers and speculators win?

Meanwhile, the sun burns down on the polar caps, and each year the beaches erode a little more, exposing tree roots and scouring away the marshes as the sea gets a little higher. We watch the weather change, the tides inch up, and wonder if in the end Mother Nature or Turtle Mother might not have the last word.

# Contents

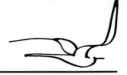

# Introduction
## to the 1988 Edition

Where there is a seashore, there is a Living Dock. When this book was first published it was called *The Living Dock at Panacea,* but it could almost have been called the Living Dock at Cape Cod, or Hilton Head Island, Key West or even San Francisco. A dock is always a place where people can step out, stand over the water and satisfy an inherent need for serenity as they watch the currents sweep by, mesmerized at the ebb of tide and the flow of life.

The composition of species changes subtly from place to place, but the essence and wonders of aquatic life remain the same. North of New Jersey there will be blue mussels with a matrix of creatures living in the tangle of byssus threads. In the Florida Keys one might find encrusting corals and brightly colored tropical fish darting among the pilings. Along a northern coast it might be a striped bass that snatches your float under, or a codfish might take your hook. In the south it could be a spotted weakfish, whiting or Spanish mackerel. But in any of these areas, kids dangle chicken necks and victoriously pull in crabs. Docks are places where earliest childhood memories are made, mine being the big pier in Coney Island.

Since this book was first written, the original dock at Panacea, Florida, has been rebuilt three times. First after Hurricane Agnes in 1972, then after Hurricane Elena in 1985, and again six weeks after Hurricane Kate. The last time, the living dock was shattered to splinters, totally bent beyond repair. Looking at the twisted wreckage the storm left behind, I knew that I could scarcely stand the expense of putting another dock back down. You can't insure a dock, and I was still making payments to the Small Business Administration from the first hurricane. But I had to have a dock. So now there is a new dock, financially encumbered but there. The barnacles and shipworms don't mind the debt.

# THE LIVING DOCK

A generation ago another man had a similar dock, which he used to explore the world of the Gulf coast in a different way. And thanks to his family, I have been able to use drawings and paintings from the works of Walter Inglis Anderson to illustrate this edition of *The Living Dock*.

I heard about this man about a decade ago when I was working on a book about shrimping. One day I was packing my suitcase, getting ready to go to Mississippi to catch opening day of the fishing season. All the shrimpers in Panacea were heading west, to be in Pascagoula when the bay opened so they could partake of the rich harvest of brown shrimp. The television was mindlessly on in the background, but I paid it no attention.

Then a film called "The Islander" came on, about an artist-recluse named Walter Anderson who had lived in Ocean Springs, Mississippi and created thousands upon thousands of water colors and pen-and-ink drawings of every imaginable creature that swam the water, flew the skies and crawled upon the land. Packing forgotten, I watched with fascination the story of his life. Professionally trained in New York, Philadelphia and Paris, he nevertheless spent most of his adult life in Ocean Springs. He frequently rowed fourteen miles out to Horn Island, one of the narrow barrier islands within sight of his home on the Gulf coast, to live as a hermit and paint.

He was considered an eccentric. People often saw him high in the branches of an oak tree in downtown Ocean Springs, sketching blue jays in their nests, and they would laugh. For Anderson, trying to make a living as an artist, the professional isolation of a small southern town and his inability to support his family was often more than he could handle. He had bouts with mental illness. After a petition to paint over his murals in the community center was circulated, he became increasingly reclusive. The seclusion of Horn Island was his salvation. He died in 1965 virtually unknown.

Among other subjects, Anderson painted the obscure creatures that we at Gulf Specimen Company sell to laboratories and schools. And he breathed life into them. His brush strokes captured the essence of wind-tortured, salt-dwarfed coastal pines, the radiance of sunrise, the glow of colors, the spirit and energy of all living things. I knew I would have to see his work firsthand, for we both actively appreciated the complexities and beauty of the Gulf of Mexico, were drawn to its life and its mysteries. Years ago I would have been amazed at the coincidence—that I was leaving for Mississippi as I happened to see the TV show. But by then I knew there are no coincidences—it is all part of one connected cosmic tapestry.

I spent the next week at sea, squandering life on the decks of shrimp boats as I culled off the trash fish. The nets came up and the nets went down and the baskets of shrimp rose higher. Sleeping by day, shrimping by night, I scribbled notes as I could. It was hot, grueling, sweaty work.

When we anchored off the Mississippi barrier islands, I thought about the hermit artist Walter Anderson. He had lived as part of the environment on Horn Island, devoured by sand gnats, yellow flies and bugs. He withstood the freezing north winds and the broiling heat, and became part of the seasons of the sea. No doubt the yellow flies biting me were descendents of the ones that had tormented him.

After the boat docked in Pascagoula, I caught a ride to Ocean Springs and turned off onto a road paved with broken seashells to the Anderson complex. Time seemed to freeze; instantly the subdivisions and carscapes of an exploding sunbelt town were left behind, and the Old South returned. It was as wooded as my home in Panacea. Jutting out over the marsh from a shoreline covered by massive slash pines and live oaks draped with moss was his family's living dock. It was from this dock, with its old silver planking, that Walter Anderson embarked in his small skiff, crossing the Mississippi Sound to Horn Island. My affinity for the man and the place grew.

After days at sea I suppose I looked a little wild, and it took some persistence before I convinced Sissy, Anderson's widow, and her daughter Mary, to let me see his work. Finally Mary led me to his little cottage beneath the moss-draped oaks of the coastal forest, not far from the dock. In the last years of his life, he allowed no one into his cottage. After he died his family opened the doors and gazed with wonder and surprise at the vast multitude of water colors, pen-and-ink drawings, block prints, wood carvings and sculptures he had left. There were over thirty thousand pieces. The amount of work produced by one man was mind-boggling.

The moment I stepped inside the cottage, I felt the richness of the place and his overpowering presence. One room was totally empty of furnishings, but every inch of the walls was covered with murals. I stood there reeling beneath the power of his artwork and vision—for there, from floor to ceiling, was a recreation of life on earth. It was so overpowering I had to sit down, to focus on it one piece at a time in an effort to take it all in.

The essence of luminous butterflies radiated out from the spirit world. A deer stood amidst the foliage, looking back at me with bland startled eyes, cranes rose up and an osprey hovered. Creatures of the night moved amidst the pitcher plants in the Mississippi bogs. I stared in awe at this tapestry of life, a three-dimensional world captured in two dimensions, with birds in the sky, turtles crawling along the ground. The shaft of afternoon light that came through the windows brought out the vivid colors of the Virginia Creeper vines, the waxy leaves of palmetto, flowers, ferns and marsh grass, all woven into one. Only the sounds of real flesh-and-blood birds calling from the real woodlands outside the windows broke the silence.

Stars burned down from the top of the walls, and an alligator crawled along the baseboard—not as a reptile accurately depicted but as the dragon from one's childhood, with great rounded eyes and shimmering scales. It crawled not from the swamps but from the mind, with all its fantastic dragonlike qualities. And skulking in the bushes, as only a feline can skulk, was a black house cat, prowling on its mission of the hunt—its green eyes aglow.

And there, in the midst of it all, was a forest spirit, a wood nymph or river goddess in human form, regarding the creatures about her with love and adoration. Tributaries and creeks branched from her head in shimmering splendor. Her face bore a wide humorous expression. She was the goddess of the forest, looking at us all, our egos and follies. I wondered when and where Walter Anderson had met this spirit, for while I hadn't seen her directly, I had felt her presence in the darkest woods and along enchanting streams in moss-grown brooks.

Slowly my eyes made their way up to the ceiling, and there radiating out in great spiraling form was an enormous pink and red flower. It was the key to all life on earth and in the universe for that matter. Awed by it all, I lay on the paint-splattered floor, gazing up, repeating the words he had scribed on the back of a drawing of zinnias left on the mantelpiece:

Oh, Zinnias!
Most explosive and illuminating of flowers
Summation of all flowers
Essence of eccentric form!
Essence of concentric form!

It was the yin and yang, the positive and the negative, the two opposing forces of nature that made the great wheel of life go round.

Anderson saw what other visionaries see, what any shaman or Zen master knows, and he tried to capture the flow of life as it spiraled its way into each and every living form. He captured the "treeness" of trees, the "crabness" of crabs and the "catness" of cats. He was fascinated with patterns of nature, the repetitive forms of convergent evolution, the symmetry of crabs, the motion of birds and the spirals of conch shells. Anderson was so enamored with energy that he lashed himself to a tree and rode out a terrifying hurricane to better understand the energy of the swirling forces of clouds, wind and water.

Suddenly I felt a great kinship with this man, for both of us shared the need to understand how life is put together. We needed to know what makes a squirrel a squirrel, a cowfish a cowfish, or a beetle a beetle. And both of us had looked for it in the same place—the Gulf Coast.

The energy in that room was so intense that I could look at it for only a short while before it became overpowering. Mary Anderson Pickard led me to an old building where many of his water colors and paintings were stored. She let me go through the conservation boxes filled with his drawings. Mary had spent years cataloging and publicizing them so others could see them. There were stacks upon stacks of folders, each marked, "insects" or "sea shells" or "turtles," "frogs," "snakes," "birds," "marsh grass" and "trees." He left nothing out.

When he worked, his sketches rained down like snowflakes, sometimes completed, often only half drawn, half colored. He was like evolution itself—experimenting with species, picking them up, developing them, throwing them down and starting again, and again and again. They were a meditation technique, a path to understanding. The sketches themselves didn't seem to matter to Walter Anderson. Many were stained and ruined, some were used to light fires, to stuff into screens to keep the bugs from eating him.

I looked through the folders until I was dazed, three thousand 8 1/2" x 11" fragile drawings on flimsy paper, yellowing with age. Just when I thought I couldn't take another, something new would jump out and make me stand back with awe. He captured the humorous mannerisms of animals—for example, how a preying mantis thrusts its butt into the air as it walks about on spindly legs and stares at the world through bulbous eyes.

Anderson drew sea hares copulating, pelicans in flight. I'm convinced that if he didn't draw it, it was because he never saw it. Stingrays, sharks, cowfish and batfish, they were all there. When he painted oak trees, like the ones that surrounded his cottage, the boughs and trunks twisted and reached up like living things—almost as if their growth were filmed by time-lapse photography. Bark fascinated him. He loved its reptilian scaly nature. He could feel the awesome patterns of wood, their flowing lines paralleling the waves rolling up on the beach.

One by one I went through his pictures of sargassum weed, blue crabs, portunid crabs, calico crabs, nudibranchs until I came to a simple and largely unintelligible sketch mislabled "worms." For it was more words than drawings. Anderson had attempted the impossible—to capture the flashing, iridescent colors of a watery ctenophore jellyfish that I attempt to describe in this book. Like the Tao itself, of which Lao-tzu wrote, no words can convey the shimmering beauty of this watery jellyfish. And no brush can render the pulsating colors as they flicker and fracture in the sunlight into prismatic blends.

He resorted to words, perhaps as a study for later attempts, hastily scribbling down the hues—the reds, pinkish reds, purples, yellows, blues, blushing greens that instantly appeared and vanished, to be replaced by another hue until the words ran off the page and I broke into laughter. When I saw Mary's puzzled expression, I explained that you had to see these transparent manifestations of living water, shimmering in the sunlight; to try to capture them in anything but their living form was folly.

The way he studied form was to draw it again and again and again, getting it "right" for the twentieth time. There was a whole folder of octopuses, a single specimen he found on the shore, each painting capturing a different aspect of its movement or color, a new position, a better view of its cold, brooding eyes or the sucker discs on its tentacles. He drew the big swamp lubber grasshoppers from every angle, turning them over and over, pulling out their legs when dead, extending their wings, essentially dissecting them visually, to see what they were made of.

To Walter Anderson death was a part of life, not to be shunned but scrutinized and absorbed. He didn't draw his subjects to hang on people's walls; instead he drew in an effort to understand what life was and how it departed. He would find a dead bird or a rotting porpoise at tideline, then convey it on paper as a powerful still life, bringing the message that no matter what your form may be, you can be certain of death—and perhaps even of ending up as detritus on the beach.

Yet in his *Horn Island Logs* he wrote of his despair at seeing some of his models perish from the elements—the pet duck, the baby raccoon. Sometimes it seemed that he sucked the life energies out of the creatures he drew in order to render them on paper.

After I went through the drawings Mary said, "I'm so glad you like them, especially the invertebrates. Everyone gets excited over the bird pictures or the trees or landscapes. But so far you're the only one who has understood his invertebrates." And there, too, I felt a kinship, for so few people try to understand these bizarre forms of life that swam in ancient seas long before there were human beings.

Anderson never shied away from the challenge. He soaked up life and emptied it back on paper, capturing its essence, the motion and form of every living thing. I was especially taken by his drawings of shrimp, since I had been handling thousands of them in my week at sea. No one had ever captured their frenetic, hysterical qualities before. They looked as if they were about to jump off the paper. In some cases, the incomplete drawings and studies say more than the finished products about how he saw life.

It didn't seem to matter to Walter Anderson that he was unknown and undiscovered in his lifetime. It was the work and the creative spirit that mattered. All that is left of his ten-and-a-half-foot wood carving of "Father Mississippi," is a faded snapshot— a river god with a blue flowing beard, tributaries branching from his head and a deer at his side. He carved it sometime in the 1950s, and it stood in the woods beside his house, where the coastal birds and animals lived around it. It deteriorated over the years, and when Anderson died, only the deer remained.

In addition to the paintings and artwork he left behind, there were log books filled with thousands of pages of observations about the beauty of the Mississippi coast. He wrote about the river, and its eternal trek to the sea, and how it built the barrier islands. The poetry of the ocean and the forms encountered there seemed to be enough for him; but even so he paid an enormous personal price. As he wrote in his *Horn Island Logs,* "Those who have identified with Nature must pay the consequences."

# 1     The Living Dock

I sat on my dock in Panacea watching the tide wash out of the bay, carrying wisps of flashing jellyfish and floating rafts of green sea grasses. I would have liked to just sit there indefinitely and do nothing, absolutely nothing but watch the afternoon light bathe the towering pines that crowded around the shoreline as the vast stretches of marshland became silhouetted against the land and the silvery waters of Dickerson Bay.

But I had work to do. I had come down to the dock not to enjoy the afternoon, nor to watch the silver flashes of menhaden minnows feeding on the bushy pink hydroids that were attached to the floating boat stalls. I was on a mission: I needed a dozen large barnacles. They had to be at least a half inch in length, so I looked among the assorted ropes and strings of oyster shells that were nailed to the dock and hung in the water.

But the plantlike growths of hydroids were now so thick and bushy that nearly all the barnacles were obscured beneath them. The strings of oyster shells that we had planted only a year ago were now so totally massed over with sponges and sea squirts that to find decent-sized barnacles I would have to wait until the tide ebbed enough for me to wade among the pilings, where I knew that big

*9*

healthy barnacles could be found growing among the clumps of bearded gray oysters.

Off in the distance, a crab boat was moving slowly up the bay, and even though it was almost a half mile away, I could see that it was loaded with crab traps. It was headed for the Rock Landing Dock, where most of the crabbers of Panacea tied up. Panacea, in northwest Florida, was a little fishing and crabbing village. The crabbers were bringing in their old traps for reconditioning, traps that were probably rotten, eaten away by salt water, and covered with growths that made them too heavy to lift. The crabbers salvaged what they could and threw away the rest.

The corks and ropes were likely to be massed over with big hefty ivory barnacles, *Balanus eberneus,* which would be larger and more desirable than any I could find growing on my dock. Maybe I wouldn't have to wait for the tide after all. I could have them packed up and ready to mail in just a few minutes.

I hurried down to my truck and drove to the Rock Landing Dock in time to see the crabbers coming in through the last channel markers. The Rock Landing Dock was a study unto itself, with all the old boats tied up to its shabby, worm-eaten pilings. The dock had been rammed by boats, abused and misused, and much of the planking was missing, but it was the only public dock in town. It was a real fisherman's dock; you could tell that not just by the fishing boats tied there, but by the way it smelled. It *smelled* like a fisherman's dock, from rotten crab bait that had been left and forgotten and from shrimp juices that had soaked into the weathered, half-rotted boards. There were great piles of discarded, rotten nets, old fish boxes, rusted chains, crab traps that would never be used again, and general clutter left by the people who fished off the dock. Anyone who liked organization would not have liked the Rock Landing Dock.

As the boat drew near, I recognized the crabbers: John Henry Williams and his deck hand, Buck Spivey. Both were wearing greasy coveralls that looked as if they hadn't been washed in ages. Buck had on boots and a khaki shirt with buttons missing, and his hair was disarrayed and flaked with fish scales from shaking crabs and decayed bait out of their traps. The boat itself was an old, battered cypress skiff powered by a decrepit five-year-old outboard motor. I waved as they came up to the dock.

"How's crabbing?" I asked.

"We ain't done nothing. We just getting these old traps up from the bay, going to clean up the corks and lines," said John Henry, grinning and showing his missing front tooth. "With the price of corks nowadays, a man's got to if he's going to make out."

"They sure are grown over," I agreed, looking down into the boat where the crab corks and lines were massed with huge fat clusters of barnacles. I could see them opening and closing their little trap doors, their feathery legs protruding, and when Buck cut the motor off, I could hear them making a low bubbling sound.

"Look at this son of a bitch," Buck drawled, lifting a crab cork and line that was a solid mass of barnacles and hydroids. "This line must weigh thirty pounds or more! Why, they had the corks pulled so low in the water we could just barely see them sticking up. You always ex-peer-minting up there, why don't you come up with some way to keep these barnacles off our crab corks?"

"I'm trying to grow them, not get rid of them," I said, laughing. "You want to sell those corks and lines? I tell you what, I'll pay you the price of a new set if you let me have those."

John Henry looked at me incredulously. "Hell yes, I ain't about to turn down a deal like that. We was fixing to recondition them, let them dry out and beat the barnacles off later. I figure I got about a dollar in each one of these corks and lines. If you pay that, you got a deal. How many do you want?"

"All you got."

"What in the hell are you going to do with them?" demanded Buck.

"Sell them. I'm going to tie them off my dock and when people want barnacles, we'll have them to ship."

"You sell barnacles?" asked Buck, astonished. "That's the craziest thing I ever heard of."

"Hell yes, he sells barnacles," the other fisherman said. "He sells horseshoe crabs and worms and all kinds of things. Why, I remember a few years back, old Jack here went out with us and was getting those big red slimy worms off our traps when we was working down at the lighthouse. You ain't never seen a body get so excited over worms," he laughed. "You ought to see the way he took on."

"You can have every one of them nasty-looking things," said Buck, shaking his head. "I hate to handle those slimy bastards. It gives me the creeps. Leon Crum can have 'em, too. He told me to save some next time I went out, and I told him I ain't got a bit of interest in messing with them nasty old worms."

John Henry grinned, showing the gap in his teeth, and shook his head. "The boy's got a weird business—that's for sure. Now, where do you want these corks? You gonna pick 'em up or what?"

"Meet me down at my dock," I said, pointing up toward the head of Dickerson Bay. "I've got to tie them off our boat stalls so they won't die. Some of them look like they've been in the sun too long already."

After unloading the lines the two fishermen came back to the laboratory with me to collect their pay. While I went into the office to cut their check, they wandered around the tanks where Leon was putting specimens into plastic bags and then packing them into Styrofoam boxes. He looked at one of the strings of barnacles we had brought in to fill that afternoon's orders and held it up proudly. "They sure are fine. I almost hate to ship barnacles like these 'cause our customers will start bitching when we send them smaller ones."

"I'll be damned if I can see why anyone would want to study a barnacle," John Henry said.

"'Cause they ain't got no barnacles up in Illinois. They ain't got no ocean. If these professors want to study barnacles, and they got the money, we'll damn sure ship 'em."

"Well I sure wish they'd study a way to keep them off the bottom of my boat. Near about as fast as we scrape them off they grow back again. You like working at this, Leon?" asked John Henry.

"I love it. There ain't no other business in the world like it . . .

It kind of grows on you after a while."

Buck was wandering around the lab looking into tanks. "How much you pay for horseshoe crabs?" he asked, watching Doug and Edward pull crabs out of a large tank and pack them into boxes filled with Spanish moss for shipment.

"Sometimes we pay a dollar apiece!" Leon replied proudly.

"Goddamn, then I've throwed a million dollars worth overboard," said John Henry, his mouth open in surprise. "Many a time we've struck Shell Point Reef and them nasty bastards got all tangled up in our nets and we'd beat their heads in just to get rid of them. And all this time I was throwing all that money away. How many do you want?"

"None right now," Leon replied. "We can get plenty of them. But we buy just about every one we can get in the wintertime, when they're scarce. Right now almost all the schools are closed for the summer, but you just wait. We'll be buying all kinds of things in the winter."

"I'll say," grumbled Buck. "Jack just bought all them barnacles from us. Bet he'll sell them for a hundred dollars when he ain't paid us but a dollar apiece."

"I don't grudge no man from making a living, and them crab corks weren't worth nothing to us no way," John Henry replied. "If that trashy mess can do some good somewhere with kids studying them, let 'em have it. I don't care what Jack makes on it."

"I'll tell you something else, Buck," said Leon in a defensive tone. "Look around you. It takes some kind of money to operate a place like this. A lot of this stuff they use in research. I bet you didn't know that that old stinging grass that grows all over the crab traps—they call it hydroids—is being used to cure cancer. We ship it out to labs all over the world. We even sent some horseshoe crabs to Israel last week. Gulf Specimen Company is going to put Panacea on the map one day."

Leon had been with me long enough to see the company rise from a dream into a reality. He could remember when it was simply myself and an airedale, an old battered station wagon and a dip net;

in a few years it had evolved into a real organization with a laboratory with running sea water, tanks and filters, a museum of preserved specimens, offices, and a few blocks away, a large fine dock with motorboats and a small shrimp trawler named *Penaeus*. We had a staff of three full-time employees who worked in the field, and a secretary who handled the orders and struggled with the books. It was a viable, functioning enterprise, and although it wasn't profitable yet, we were hopeful that someday it would be.

Gulf Specimen Company was a business born of man's fascination with the sea, his interest in exploring its waters and learning about the strange creatures that live down in the depths, sharing his world. Our business depended upon the ripeness of sea-urchin eggs, the movements and migrations of sea hares, the prevalence of sharks, and the spawning of horseshoe crabs on the sandy beaches. We gathered our inventory from the marshlands, and our trade secrets were how to dig up lugworms from mud flats and how to find brachiopods by their knifelike slits in the grass beds. John Henry was right: it was a strange business of shrimp nets and dredges and shipping live sea creatures so they would arrive at their destinations in good condition.

Sometimes I was overwhelmed by the technology involved in getting these creatures out of the sea and into the laboratory. As we pried a quahog clam out of the ground and put it into our bucket, I realized that we were operating in a most primitive fashion, probably not much different from the way in which the Indians hunted them more than five thousand years ago, when they lived along the seashore and piled their shells up into great mounds that were still seen in the marshes around Panacea.

But these same clams, *Mercenaria campechiensis*, that we dug up on the tide flats were packed into Styrofoam boxes and jetted across the skies, ending up in laboratories where some of the most sophisticated equipment and technology that mankind has yet achieved would be used in studying their every aspect: their molecular structures, their weights and patterns, their ability to withstand carcinogens. Physiologists removed their nerves and hearts and recorded their nervous impulses with electrodes and oscilloscopes. And so the clams, and the lowly brachiopods and other creatures that have existed for hundreds of millions of years that we gathered from the tide flats, did not go to feed man's body, but they fed his mind.

Mary Ellen, my secretary, came out with a telephone message. "While you were down at the dock a customer called in. He wants some crab I've never heard of, and he wants it gravid."

I looked at the note. "Urgent. Two dozen gravid *Neopanope texana texana* needed before next week. Specimens must have egg sponges or they are not acceptable."

"He wants to know how much you'll charge," Mary Ellen said, waiting as I studied the note.

"Charge? I don't know; I'm not even certain which crab he's talking about. I think it's the little xanthid crabs that live in sponges on our dock. I don't even know if they're gravid. Let me go look, and if I find any you can call him back," I said, happy to return to the dock.

*Neopanope texana texana* is an obscure little crab that lives under rocks, among oysters, in sponges, and practically anyplace that it can find shelter. It's nowhere near as impressive as its name. It doesn't have the size or pinching power of the blue crab, nor the big grinding claws of its close cousin the stone crab, nor the flashy colors of the portunid crab. In fact, there really isn't much to recommend these drab little crabs, which are barely a half inch across, and there is even less that one can say about them.

As I sat upon one of my floating docks, tearing sponges and clumps of hydroids from beneath the Styrofoam bottoms of the boat stalls, I picked out the little mud crabs and put them into a bucket. So far I had gone through two large clumps of sponge, but hadn't found even one gravid female. Like all other crabs, she carries a black spongy mass of eggs under her apron until her larvae develop and are released into the water. I knew I had seen those little crabs gravid from time to time over the years, but they were so small and insignificant that I really hadn't paid much attention.

There weren't as many as I'd expected; I found only a few

dozen, which I put aside to feed to the dwarf octopuses back at the lab. There were plenty of females. In fact, they seemed to outnumber the males greatly in the sponge colonies. You could spot the difference at a glance by examining its apron, which is the plate that covers the underside. The females had broad aprons that rounded to a point, and the males had long pointed skinny aprons. Maybe the crabs had already spawned. Or maybe they would develop and produce sponges later on in the fall. But whatever had happened, or was going to happen, didn't matter. It was unlikely that we would be able to produce gravid *Neopanope texana texana* by next week, so the fifty-dollar sale (that was what I had decided to charge for a special collection) was lost.

Why couldn't that professor in Maryland want some of the ctenophore jellyfish that were drifting by the dock, swept along by the currents, flickering and glittering and catching the rays of the sunlight? Why didn't he want some of the marine water striders that danced over the surface? Or why not any of the other creatures that lived attached to my great and wonderful dock that stretched out over the mud flats and stood rooted in the water column?

I enjoyed my dock, gravid crabs or not. Sometimes I failed to bring back the animals I sought, but often I came back with my buckets brimming with spider crabs, blennies, barnacles, and other creatures. I had to admit that it was the best investment I had ever made—it afforded me hours of endless pleasure, and provided me with an income at the same time. We could sell the hydroids that blossomed from the pilings along with the stinging nettle jellyfish that drifted by with the currents. Often I would sit on the dock with Anne, my wife, watching the tide come in and cover up the mud flats and rise up into the marshlands, where we would see periwinkle snails crawl up the grass.

Docks are wonderful things, although they really go unheralded. Stories have been written and songs have been sung about ships and mighty vessels that have gone forth to conquer the world, but no one has really glorified a dock. It stands there mightily against wave and wind, and puts up with the woodboring creatures that chew at its pilings, which stand firmly rooted in the mud. A dock fills a need of mankind: it provides a bridge between land and sea.

When you go out onto a dock you leave the problems of land behind you, yet you are not involved with the problems of the sea. You don't have to worry about boats breaking down or whether you're going to run out of fuel or whether your dock is going to spring a leak. If the weather is bad, you just turn around and go back to shore. Somehow, just standing there looking down into the water makes trouble fade off into the background. All your problems are left on the dry land while you have almost all the security of land beneath your feet and all the openness of the sea before you.

From the moment the first piling is driven down, a dock begins to become part of its surroundings. At first, creosote leaches out of the pilings and repels the planktonic larvae that try to settle, but sooner or later that first thin layer of white barnacles forms on the surface. It matters not what you put on the piling or hard surface to repel the larvae; the sea can wait. All you buy is time, one year, two, three, maybe even five or ten years, when you use the most expensive and toxic antifoulant substances. Sooner or later the sea takes away their potency.

A bacterial coating covers the surface and mixes with mud that is suspended in the currents, forming a substrate for filamentous green algae. Then the cyprid larvae of barnacles touch it and find that it is good. Within a few months the veliger larvae of oysters are no longer repelled when they touch it, and the planula larvae of hydroids start to bloom into bushy flowering pink growths. The dead wood that you have jetted into the sand and mud begins to live again. The sea gives it life, and life begets life. Before long a distinct zonation of living forms develops on the dock, which is stationary and affected by the tides. The pilings that are closest to the shore in the shallowest reaches of the bay are covered and uncovered by the tides, day in and day out. All but the bottom few inches of the pilings are exposed to air, to the world of harsh icy winds, which dry them, to the broiling heat of summer, to the splatterings of rain.

Only a few hardy oysters and perhaps some barnacles can live on these first few pilings, but ten feet farther out, on the next pair of pilings, there is a little more growth. The oysters aren't tiny flattened shells that barely survive; they are a little bigger, and there are some ivory barnacles because these pilings normally stand in a foot of water. By the time you reach the end of the dock, where the pilings are rooted under four or five feet of water, seldom exposed to the elements of air and dry land, there is a rich growth of soft-bodied hydroids, sponges, and sea squirts, along with the more robust ivory barnacles and big fat juicy oysters.

Some of these soft creatures cannot survive desiccation for more than a few minutes and can grow only if they are constantly covered with water. In the cold winter, when the tides are low and the pilings and mud flats exposed, the freezing wind blasts down from the north and dries out the soft-bodied creatures, and they perish. Only the oysters and barnacles, secure in their shells, survive and wait for the tide to return.

Floating docks, which rise and fall with the tide, provide a striking contrast to the stationary structures. Originally I had built them for convenience, so that I could get into my boat in high or low tide, but I soon learned they had other advantages. They enabled us to practically walk on the surface of the water, to sit down and be only a few inches away from the creatures that swam around the pilings.

When the tide was starting to swell into the little marshy bay and covered the oysters and barnacles on the pilings, the floating docks rose. Attached to separate pilings by steel rings, they moved vertically as the tide moved. When the bay was swollen during a spring tide or a storm, they stood floating like battleships high up on the waves, and I could step from the stationary dock onto the floating Styrofoam dock with no trouble at all. But when the moon was full and all the waters were sucked out of the bay, or the powerful north wind pushed the waters out, exposing the mud flats, the floating dock sank far below the oyster and barnacle zones on the pilings. In fact, there were times when the winter wind shoved all the water out of the bay except for a narrow channel, and the floating dock sat squarely on the bottom like some sort of stranded sea monster bearded with huge amounts of tufted oysters, barnacles, and other fouling growths.

The floating docks are much better places for delicate fouling organisms to live, because these great flat surfaces of Styrofoam and wood are almost never exposed to the air. They sit on the water like the hull of a boat, so creatures of all sorts find their home in the Styrofoam. Most simply attach, but stone crabs like to dig out holes and tear out burrows in the bottoms of the floats; old Styrofoam boat stalls become honeycombed with small crabs after a few years and then numerous other creatures move into the silt-lined burrows.

After the floating docks are assembled and strapped and nailed into place, it seems little more than a few weeks before they are sprouting a growth of hydroids. Before long, a thin layer of barnacles forms. You can lie upon the dock and watch the barnacles fling out their delicate feathery legs and sweep the water for plankton, then draw them back into their little armored shells. After a time you may notice that patches of serpulid worms have started growing on the oysters, secreting their tiny white twisting tubes of lime. It seems that almost overnight a crumb of bread sponge has sprouted, and magically the sponges become overgrown with little clusters of pink sea anemones, which continually bud off more anemones. When you build a dock, especially a floating dock, you create an ecosystem crawling with mud crabs, filled with pink bristleworms, and harboring a multitude of small shrimp and tiny crustaceans.

After watching a fouling colony for a time, you might think all these creatures appear by spontaneous generation, because it takes place so very rapidly. A layer of sponge forms sooner or later, and when you examine it you'll find all sorts of life hiding down in its intricate vascular canals and partaking of foodbearing currents that are pumped through the colonies. Sometimes there are brittlestars, sometimes pistol shrimp or little white amphipods that have become highly modified both in form and in color, depending upon the host colony to provide assistance in the reproductive capacities of its "guests" by pumping out sperm, eggs, and larvae.

I doubt that any habitat in the world has the competition for space that a fouling community has. Every inch on a dock is colonized by barnacles, or oysters, or hydroids, or sponges, each trying to push the other out or coat it over. Sometimes they succeed and one group will supplant everything else, other times no one succeeds and the dock is choked with life until the environment changes by season or by weather. The old is scrubbed off or dies and then life begins anew.

It is well known to fishermen that a large variety of edible fish like to prowl around docks. Some, like trout and redfish, come to forage; others, like mullet, which are detritus feeders, often just swim around docks, perhaps feeding on the rich organic muds that are trapped by the bushy hydroid colonies or perhaps seeking protection from predators. Many nights fishermen in Panacea would come down to the dock and stand with their cast nets poised, ready to throw when they saw a ripple. And often when the net came crashing down into the water, a few mullet would come flopping and jumping up. If they used a small-mesh cast net in the late summer and fall, when the shrimp were migrating out of the channel, it wasn't hard to get up five or ten pounds in an hour. And of course there was always the good old hook-and-line method, which worked for those who were patient.

Nicholas was patient—more patient, in fact, than anyone else I knew. He would sit on the dock hour after hour and fish, and if there were any fish to be caught, he would catch them. Hook-and-line fishing never appealed to me, perhaps because I fidget too much and don't have the patience to sit still, waiting for something to tug at my line. But I liked to watch Nick fish. He was a distraction for me. I often sat and talked to him for hours while he expertly dabbled his line until suddenly his cork shot down under the water; he would pull back, and then up would come the glittering silver yellow tail, or, even better, a speckled trout or flounder.

On the particular day that I was looking for xanthid crabs and watching the tide fall and listening to the porpoises in the bay go "phoof" as they spouted water up through their blowholes, Nick came down to the dock to fish.

"What you looking for?" he asked, seeing me with a big pile of torn-up sponges.

"Little mud crabs, with eggs on them."

"Eggs on them, huh?"

"Eggs."

"Find any?"

"No."

Nick reached into his plastic bag of thawing shrimp, expertly slipped a hook up along the back, and tossed his line into the water. For a moment the line sat there limply, then suddenly the cork began to twitch. Nick had a set expression on his old weathered face, but a grin started to form as the cork began to move, and then he snatched the line and reeled in. When the fish broke the surface we could see the dazzling golden-and-red-body channel bass that everyone referred to as "redfish."

"Nice little puppy red," Nick said, holding it up proudly, and in a moment he was fishing again. A few minutes later he had three reds, two croakers, and a speckled trout on his line. The fish were really biting in the falling tide. I watched him shove the steel nail of the stringer through the silvery fish's gills and into its mouth, and then lower it into the water along with the other fish.

I felt sorry for them. I thought that those fish were probably in pain. Maybe their nervous systems were crying out in agony as they were tied to the strings in the water, yet they had no ability to cry out. And because they couldn't express pain, nobody sympathized with them.

I didn't voice my thoughts to Nick—instead we talked about the hot weather, the politics of the little town of Panacea, and the fishing. I had given up hope of finding any gravid xanthid crabs for the day, but maybe there would be some offshore in sponges that we dredged up in deep water. I decided I might as well do some other work. I had been assisting some students at Florida State University in a research project to examine the gut contents of various commercial and sports fish. This helped establish a broad understanding of the food chains, showing which animal fed upon which.

"Nick, how about letting me dress some of these fish for you? I want to examine their guts and see what they've been eating."

"Sure, you can take some home for dinner if you want to. It looks like I'm going to have plenty."

"No, thanks, I'll take a raincheck. I'll just gut them for you."

Using his razor-sharp pocketknife, I slit open the redfish's belly, exposing the pink intestines, and felt around until I located its bulging stomach. Upon slitting it open, I found a small blue crab. It was pulverized from those powerful jaws, but very fresh-looking, as if the fish had just caught it a few minutes ago. Then I cut the trout open, and in its gut were a number of grass shrimp and small brown crabs.

"Now where do you suppose he got all them little crabs and shrimp?" asked the old man when he saw me spreading out the little bodies.

"Probably right around this dock," I said. "These shrimp are the same species that we catch in the dip net right off the floating dock, and these crabs are the same ones I just picked off the floating sponges. There's thousands of them all over the dock, and fish love to eat them."

"I'll say. Since you been handing all this stuff out for things to grow on, this is one of the best docks around here. Hell, the Rock Landing Dock ain't near as good."

"It's like an artificial reef," I told him as I put the gut contents into a bottle. "All you do is put something into the sea, like old tires or cars, or pilings, and fouling growth takes it over. Then fish come to it."

"Well this ain't just a dock then," said Nick, who had put down his fishing rod and was looking over my shoulder at the next fish I was dissecting. "The whole damn thing's alive. This is a living dock!"

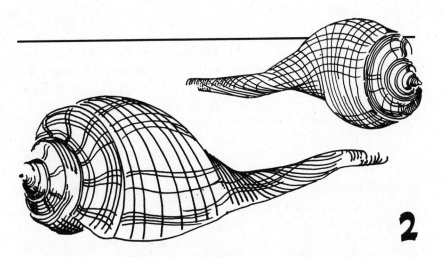

# 2

# The Live Bottoms

It was already the twenty-fifth of July; in five or six weeks, researchers would be returning to their laboratories and teachers would be starting their classes. We would be hardpressed to fill their orders unless we got out and started stocking our tanks with every conceivable creature.

What a perfect morning it was for an offshore dredging expedition. The sun was rising over the marshes of Fiddler's Point, and the waters in Dickerson Bay looked as slick as polished glass. Sitting on the end of my dock, I could see the reflection of our little shrimp trawler, *Penaeus*, along with the reflection of the wharf pilings and the underside of the dock. Within an hour or two we would be heading out of the nearly enclosed marshy little backwater, down the channel, and into the Gulf of Mexico.

While I waited for Leon, Doug, and Edward, I was engrossed in watching the hundreds of delicate gray water striders that waltzed over the surface of the water, so light that they couldn't cause a hint of a ripple. A mullet leaped out of the water and shattered the glassy stillness, and a porpoise moved lazily about the bay.

I didn't have long to wait; in a few minutes I heard the truck turn off the highway and start down to the dock. In just a moment the mood of the quiet morning had changed, and we quickly unloaded the truck, piling provisions, Styrofoam chests, and buckets and ice into our dock cart. Then we rolled it down the full three hundred feet to the end of the dock, where *Penaeus* was tied.

Edward was a shaggy-haired blond youth who was generally quiet. He didn't fare very well with the English language, but he was good at other things, like fixing diesel engines and welding. A hard worker, he had been a crew member on a big shrimp boat, and he loved the work.

Doug also loved the work. He was a studious, bearded young man, who had finished his college education in Pennsylvania and come south to find work in the profession he loved best, biology. Doug thoroughly enjoyed working with salt water aquariums, feeding the specimens and collecting them. In the few short years he had been with me, he had amassed a substantial background in practical marine biology.

Leon turned the key, and the engine grumbled as the starter turned over. "Come on, you bitch," he said. "Don't give me any stuff." And then *Penaeus* started, all her four pistons pounding happily. Black smoke blew out from the exhaust, and Leon watched the oil-pressure gauge intently as the needle climbed. "She's fine," he said, nodding in approval. "I might get someone down here to adjust that miss in the fuel pump one day, but right now she's all right."

Edward and Doug cast off, and in a moment we were moving over the calm flat waters of Dickerson Bay, leaving our dock behind. We traveled past the stilt houses along the wooded shore, and waved to fishermen who were loading their boats for their morning's work. They waved back as they stacked boxes of frozen alewives and skipjacks on their skiffs to use as bait. The wake of *Penaeus* made their boats rise and fall slightly, and the waves washed against the old, half-sunk, derelict shrimp boat that was once called *Isabel*.

We chugged past the clutter of the Rock Landing Dock, down the channel, past the beautiful green marshes that mirrored their leaves on the glassy waters. Egrets walked on the shore, and now and then I saw a number of them sitting together in a hammock on the shoreline, in their eye-catching brilliant whiteness. Overhead, a flock of wood ibis winged their way across the bay toward Fiddler's Point, where they would loaf on the salt flats.

A few minutes later we passed discarded boxes and plastic bags that were used to package frozen fish, bobbing along the surface. The fishermen used the ocean as a convenient garbage dump, and it looked it. But soon we were past the debris and heading out to sea.

Then we found ourselves in the midst of huge numbers of cannonball jellyfish, *Stomolophus meleagris*, pulsing along rapidly in the current, opening and closing their white umbrellas fringed with reddish purple skirts; the sheer white bodies stood out starkly against the dark waters.

Small silvery harvest fish, *Peprilus paru*, swam along with the big jellies, using them as protectors from any fish that might be attracted to their flashy polished bodies. But they were by no means immune to the stings of the cannonballs; when placed in a bucket with their host, they quickly perished. I sat on the bow and watched the little fish revolving around their protectors, some swimming in formation at the jellyfish's side, some at the tip, and many in the rear.

I remembered reading somewhere that harvest fish feed upon jellyfish as well as seek their protection. The brainless, rubbery white ball that continued to pulse when it was chopped into pieces would probably never know the difference. However, the protection *Stomolophus* provided must have been superb, because even thick-skinned sharks moved away when the shoals of jellyfish moved into and dominated the inshore waters. They would be seen everywhere in the bays, estuaries, and offshore waters from June through September, and there would even be a few strays in the fall. But when winter came, the cannonball jellyfish would vanish, and other species would take their place.

As we moved farther out into the Gulf the shoreline faded rapidly behind us, and soon we had passed out of the area of the cannonballs. The sea was like a mirror, and on its flat glassy surface we could see the outlines of hammocks and woodlands, but before long even these disappeared from sight. Summer is the only time of year when you can count on calm hazy seas, when often the water is so slick that there isn't a ripple miles from shore. In the winter the cold winds blow, the seas are stormy, and working out there can be miserable. When you make your living from the sea, you must know a lot about her moods and her seasons. When the March winds blow with a fury, there is no sense going out. But even in the calm summer the sea can be dangerous, because of the sudden squalls and waterspouts that brew up out of nowhere. You learn to watch the directions of the winds and the movements of the tides and currents. Most of all, you must learn to respect the sea.

As we moved slowly over the glassy calm, we came upon schools of small menhaden that were feeding on the surface. Tens of thousands of these tiny bait fish flurried wildly, breaking up the clear quiet blue water like an egg beater. It was a common sight in summer, these huge aggregations of tiny fish bunched together for safety. But in winter, when the water temperature dropped, the fish would disappear, and only after the heat had warmed up the sea in May or June would they reappear.

Above the school there would often be a number of gulls, terns, and pelicans, screaming and screeching and flapping their wings as they dove down and grabbed up mouthfuls of the silvery little menhaden. I knew that down below the school lurked much larger fish with toothy jaws, like mackerel and sharks, that struck and ate the little fish. Only when the sea was this calm did the fish come to the surface, perhaps to gulp the water that was richer in oxygen than were the depths.

They stayed together in groups, and when we moved through them, with the enormous *Penaeus*, they vanished from sight.

We came upon one particularly large patch and Edward remarked, "Boy, I sure wish I had a gill net out here. I could make some kind of money on them as crab bait."

"No hell you couldn't," said Leon, as he turned the wheel and headed due south into the Gulf of Mexico. "They're too small. Most of them were little biddy anchovies in that last patch. We passed a big school of flatbacks back there, and that would have made good crab bait. The crab houses are paying five cents a pound for them. But these ain't good for nothing."

I watched a pelican rise high into the sky and come crashing down into the midst of a school as if it had just been shot down. The water erupted with an enormous splash and the humorous-looking bird bobbled around on the sea, madly swallowing up the silvery bodies with his huge engulfing beak.

"I don't know," said Doug thoughtfully. "The pelicans think they're good for something."

Since it was such a fine day, the sea was crowded with boats—another sign of the season. In the winter when we came out here, we were the only boat for miles and miles, and it was a lonesome feeling because there was always the danger of breaking down. But as soon as the weather warmed up and the seas ran down and slicked off, the sports fishermen appeared. Every year as the population of Florida swelled, their numbers increased. Some were retired residents who lived in the expensive houses that had been built on the dredged-up canals; others were weekend tourists who migrated to Florida from nearby Georgia. They came, they enjoyed the sun and the surf, they filled their ice chests with fish, and then went home to eat them. There was nothing wrong with that, but as I watched their boats clustered around the schools of flurrying bait fish, I only wished there weren't so many of them. Just the mere rumor that the Spanish mackerel were running clogged the highways with cars, trailer boats, and motors. And when the big king mackerel moved in, the motels and restaurants were jammed and

overflowing. I had moved to Panacea years ago to avoid crowds, and now the crowds were here also.

Today the mackerel were biting, and we had to avoid running down the engrossed fishermen who were pulling their shiny lures around the edges of the baitfish flurry. They were reeling in Spanish mackerel as fast as they threw their lures out. It was always good to fish around those hordes of glittering silvery fish that swam in endless circles. They stayed at the very surface of the water, hoping to be out of the visual range of the predatory fish below, and for the most part it worked. The few that strayed off were snapped up by the jaws of the mackerel.

I loved to watch those fish. They were the same ones that hung around my dock, feeding among the hydroids. Those little minnows ravenously devoured the decaying marsh and swamp vegetation that leached out into the bays and estuaries. I was standing aboard *Penaeus* watching the little fish shower out of the water when something struck among their midst, and I knew I wasn't looking at a textbook idea of food chains with little arrows and balls showing that A eats B eats C and so on. I had seen these fish greedily gobbling down the mold, algae, and fungus grown over leaves of sweet bay, gum, live oak, and even sabal palmetto that had washed out to sea along with the decaying marsh grasses. They were transferring the energy from the sun to the green plants into protein, from the lands and rivers to the sea, and finally through mackerel into man and shark. It was an amazing and dynamic process.

Leon squinted over the vast shimmering horizon, holding his hand over his brow as he looked for the buoy. Somewhere out there was bell buoy 26, which told seagoing oil barges that they were approaching land. Now and then we would see a large barge offshore, sometimes a string of them being towed by a tugboat to the refinery at St. Marks.

At last he spotted the buoy, pointed to it, checked his compass bearings, and we continued on course. A half hour later we passed it, and steadily the engine of *Penaeus* chugged on.

Now the water around us was as blue as the sky, clear and deep, and the sportsmen were far behind us. We had left the estuarine waters where the muddy rivers drained into the sea, the shallow waters were behind us, and now we were ready to drag the deeper bottoms and bring up some of that wonderful color and diversity that we knew was down there.

"All right, let's get the dredge over the side," Leon ordered Edward. "We're fixing to get into the rocks. You hear me?

He twisted the lock on the winch, and the heavy steel drag splashed down into the sea, pulling the stainless steel cable down with it. The drums on the winch spun rapidly as the steel-webbed frame sank into the depths. The water was so clear that we could see it go a long way down, getting smaller and smaller as it disappeared into the abyss. When it finally rested on the bottom, Edward hooked the cables through a heavy hook and chain that kept them centered in the middle of the boat, while Doug hoisted the cables up with a block and tackle. It was a complicated arrangement that all single-rig shrimp boats must go through to keep the tension on their masts and davits evenly distributed; otherwise, the net or dredge would tear down the whole business. At last the block and tackle were pulled taut, and Leon pushed open the throttle and *Penaeus* began churning ahead, dragging the heavy steel dredge over the bottom.

The boat moved forward easily and steadily and Leon shook his head and said, "Damn, we ain't in rocks like I thought we were. But don't worry, we'll be getting into them in just a minute."

"We can use some of those big purple sand dollars that we find out here, and some *Astropecten* starfish too, so we're not wasting our time," I said reassuringly.

It took a long time before we really found the live bottom. *Penaeus* forged mightily ahead, pulling the heavy steel dredge with its chained bottom and thick webbing over the sand bottom. It seemed that we had pulled for miles with only steady hard tension on the cables. If we struck rock, the mast would rattle and shake and even jerk the boat backward. Now and then the rigging gave a hopeful jingle that it had struck something hard, but then it was back to a smooth pull.

Although there are rocks out in the Gulf bottoms, they are restricted to a few areas. This live bottom is a premium for fish. Sports fishing guides know where it is by instinct and years of experience, but even so, when they run out there they pull car springs and a few lengths of chain behind their speed boats until they feel it banging and clattering. The more sophisticated ones use depth recorders. Then they anchor their boats down, and sit on the rocks and catch grunts and sea bass and sometimes grouper and snapper. A man builds his whole reputation on being able to find the live bottoms, and Georgia tourists soon learn which guides can put them on location the fastest.

The rocks sit on the bottom like an oasis, surrounded by miles and miles of sandy bottoms. And like an oasis in the desert they are beautiful and serene, full of vivid color and diversity of form.

Leon finally got tired of pulling the dredge without finding the rock. "Might as well get her up," he said dispiritedly. "There's not a bit of use in dragging here. I'm fixing to run about a half mile due west of here, there's got to be rock there."

When the dredge was winched to the surface it was crammed with a variety of sand-dwelling creatures such as starfish, sand dollars, whelks, hermit crabs, and other animals that we needed. You can't find the handsome velvety-purple cake urchins, *Encope michelini,* inshore where the salinity is low. Nor can many of the vivid orange and blue starfish, *Astropecten articulatus,* with their perfect symmetric arms, be found in the inshore estuaries. We raked through the catch, breaking apart dead sand dollars, picking out the reddish purple echiuroid worms protruding their white snouts, and saved the hermit crabs that wore cloak anemones; then we culled off the rest.

Leon ran on at rapid speed and the dredge dangled back and forth from the davit, swaying with the forward motion of the boat. We had tied it down because if the boat started rocking a hundred pounds of steel frame and chain swinging madly could do a lot of damage. "You've got to watch that thing," Leon had warned us. "If she ever gets out of hand and starts swinging and you're in the way, she'll bash your brains out before you can blink an eye."

When we had moved a considerable distance, the dredge was lowered again, and this time the rigging started to vibrate and rattle. Leon grinned with satisfaction. "Now by God! Didn't I tell you we'd find it? We're going to come up with some rocks that will be monsters. I just hope we don't get one too big to lift!"

As the boat crawled forward it began to clatter and shake. The mast bent and stretched slightly and the cables were pulling hard, then coming slack. "Don't stand under that mast," Leon warned Doug, who was peering into the water at the sand dollars. "You never know . . . we might hit a rock or something and bring that mast crashing down on your head. A boat's a dangerous thing—hose cables could pop and it would cut your head off before you know it."

But *Penaeus* did none of that. She just kept clattering and banging and pulling the dredge along the rock bottom, and then suddenly the cables stretched mightily and yanked the boat backward. "Goddamn, we got a good one," yelled Leon, easing up on the throttle and taking the boat out of gear so that it would not further strain the mast. "Let's get her up."

Round and round the winch drum turned, drawing in strand after strand of steel cable, winding it tightly on the drum. Since the dredge was obviously stuck down below and wasn't going anywhere, the boat was being pulled sideways until it was right above the wedged-in dredge.

The winch started pulling slower, and the cables began turning with the greatest amount of tension on them as the dredge was fighting against the rock below. At that point two things could happen: it could snap the cables or it could break the rock and free the dredge. The boat was leaning and the davit was bending down toward the water. Then suddenly it popped up and the winch began wheeling faster, and for a moment I feared the worst, but then I saw that the cables still had weight on them. The dredge had broken free and was headed for the surface. Steel had defeated rock.

At last we could see it coming rapidly to the surface, this big steel frame with its heavy green webbing. And as it approached we could see down into its mouth, the bright yellows and reds of the sponges and corals, and the purplish whites of the encrusting algae. It was a good sight; it meant that good things were coming up.

The dredge broke the surface and the heavy bag filled with rocks and sponges sagged down in the water. Edward hurried over, crawled onto the davit and hooked a rope around the bag to choke it off, and then ran the rope through a block and looped it around the brass cathead. Then Leon started bringing the net up. In a moment the dredge was hoisted out of the water and suspended above the deck.

What a sight it was, too, with brick-red sea fans sticking out of its heavy green webbing, long orange finger sponges caught up in the chain, and rocks adorned with purple and white encrusting algae and bryozoans. With a jerk of the bag rope, the whole lot disgorged on the deck with a heavy rush that almost sounded like coal pouring down a chute. Leon hurriedly eased the dredge down onto the deck, Doug and Edward shoved it to the side, and we dived into the pile.

"God, I love this stuff," said Leon as he started moving the rocks aside. "Look for octopus. This place ought to be crawling with them, and we need one for the order that's coming up next week. Goddamn! Here's a lion's-paw scallop," he cried in excitement, holding up the magnificent shell that looked like a glorified scallop, full of reds and whites and purples. "That's a perfect one. Throw it in a bucket and keep up with it, Doug, and don't put no rocks on top of it, either. . . ."

"My God, look at this," I gasped in an awed voice as I lifted a rock out that was coated with fiery orange sponges so bright that they almost hurt my eyes. And on that same rock was a large cluster of rose-bud coral, *Phyllangia americana*, a species so beautiful that it indeed looked like little rose buds. They had a cluster of very large pink and white polyps, and in aquariums they bloomed with warty green tentacles, appearing so splendid that you just had to have one for a salt water aquarium.

Although I had seen the dredge bring up this eye-catching magnificence any number of times, there was really no getting accustomed to it. No treasure chests brimming with rubies and sapphires, diamonds and emeralds, or gold and silver could compare with it. Looking at that deck, with the emerald-green mantis shrimp scuttling and dancing among the rocks, the flaming orange sea squirts, and the algae that appeared as little glass balls of shimmering maroons attached to stalks that grew on rocks covered with white and yellow sponges, I felt that I was looking at Pharaoh's treasures.

Out of this treasure slithered brownish purple octopuses, about two inches long, stretching out their tentacles, grabbing on to the rocks and sea fans with their tiny suction cups, doing their best to escape the confusion. Rapidly we grabbed them up and put them into isolation buckets, where they spurted a cloud of ink and then settled down to hide in the rocks we had provided for them.

There were many creatures in the catch that were valuable because of their form and color. How do you begin to describe the diversity and color that comes up in a dredge? Do you talk about the orange and yellow knobby starfish, the needly pink sea urchins, and the speckled brown and white crabs? Do you first consider the large flat piece of rock in the middle that has five brick-red sea fans growing on it with their intricate branches, a specimen that would make any aquarist drool? What about the rock shrimp with its dull gray hardshelled carapace, dazzling pink and white striped legs and abdomen? There were also green mantis shrimp that live within the crevices of the rocks, and roll and jump and dance about the deck so quickly that we can barely catch them. Unless we're really careful, they can jackknife and embed their razor-sharp tail spines into our fingers.

Speckled and fluted nudibranchs, orange and black in color, and bright red finger sponges forced our attention. The eye sees all this; it strives desperately to take it all in and to hold it in its memory, but it really can't. You want to take back everything that you can, but there is only so much room in the buckets. Too many creatures produces overcrowding and then everything will die.

There were great piles of spiny pink sea urchins on the deck, some of them larger than the palm of my hand. I cracked one open and was glad to see it densely packed with yellow roe, which was delicious when spread on crackers. Sea urchins were one of our mainstays, sold to embryologists who study their eggs, but we had plenty in our tanks and could get plenty more inshore. It seemed a crime to watch Doug and Edward shoveling them over the side, but there was no room for them.

We squatted down next to the piles of rocks, picking them up and looking them over, and either deciding to keep them because they had chiton or feather-duster worms on them, or tossing them over the side. Leon put the dredge out again and joined us periodically while we culled, pointing out a piece of green and white rosebud coral that we had missed or holding up a big brown glove sponge.

We worked into the afternoon, hauling up the dredge and picking out what we needed. We were after coral, bright red sponges, yellow nudibranchs, and lion's-paw scallops. Almost everything else went over the side. We were so absorbed in the catch that we didn't notice the big storm clouds building on the horizon until we felt their first cooling breezes. The dredge had come up for the fifth time.

"We ought to start heading back," Leon said after a while. "Look over there to the east. We damn sure don't want to get caught out here. Don't *ever* let yourself get caught with one of those storm clouds between you and the hill."

All around us the sea was calm; there was hardly a breeze in the air, but the clouds were stacking up. "Do you think they'll come this way?" I asked.

"I doubt it, but you don't never know about them things. You just have to watch it."

"All right, let's make this the last drag. We've got over a hundred turkey-wing clams, and this next drag should bring up more than we need."

I went back to culling the piles of rocks. On the horizon, way out at sea, great clouds were stretching high up into the sky. Right above us the sun was shining. It was now past four o'clock and the midday heat was broiling down on our backs, so everyone took off their shirts. Only in the Gulf could you have squalls like that, where in one place the sea was slick and glassy and a few miles away it was storming and raging. It would have been better if the skies had been generally overcast; then the sun's rays wouldn't have been beating down into the buckets.

While Leon steered the *Penaeus* along, dragging the bottom, the rest of us were changing the water in the buckets to keep our catch cool and well oxygenated. The sun was so strong that even with bags of ice to cool the water, the heat quickly robbed the water of its oxygen. It only took a few minutes before our prize rhinoceros blennies, sea horses, and octopuses would come gasping to the surface. In the winter, when the water was cool, it held oxygen better and our survival rate was higher, but now there was danger of losing our catch unless we were diligent.

The breeze that blew down on us from the clouds felt delightful. Cool moist air brought relief from the scorching sun, and I stood up to stretch and glory in it. With a little luck we could keep running ahead of the rain clouds and get the benefit of their breezes. It would be a very long, hot run back to the dock, at least two and a half hours, and we would be traveling against the tide. I was glad that we had another load of rock coming up; it would help occupy our time as we chugged over the sea. Travel time goes fast when you have a big heap of interesting creatures on deck. You can examine them at your leisure, look for new animals that you might not have seen before, like small worms or amphipods that hide down in the crevices of the rocks.

I picked up a large rounded rock that was riddled with burrows and holes and overgrown with sponges. Put a rock like that in an aquarium and suddenly you'll notice that it starts to spring tiny sea anemones that you didn't spot before. Before you know it, maroon tentacles of feather-duster worms spread out like a fan from their tubes that are bored down deep into the rock. If you were to take this magnificent thing with its thin layer of purple encrusting sponges and patches of white and red algae, and crush it with a hammer, you would find that the inside of the rock was alive and riddled with peanut worms, polychaetes, and boring mussels.

*Lithophaga bisulcata*, the boring mussel, makes its long pointed burrows in the rocks by chemical means, secreting an acid that dissolves away the limerock. The mussel rotates its shell as it bores a hole deeper and deeper into the rock. A thick leathery covering, called a periostracum, protects the shell from its own secretions. While the burrowing mechanisms of the mussel are understood well enough, no one really knows how the soft-bodied little peanut worms that one finds in the rocks can wallow out such neat little holes for themselves in such a hard substrate. Perhaps this process too is chemical. It seems unlikely that it could be mechanical.

I was pleased to see that a number of turkey-wing clams, *Arca zebra*, were growing on the rocks. These little bivalves are interesting because they secrete a byssus that enables them to attach to rocks. The secretion is a soft green glue that hardens into a fibrous substance, and no matter how you try, you can't tear or break it. When we wrenched the clams off the rocks, the little green holdfasts were pulled out from between the valves, but they wouldn't break. A university that was trying to develop a dental glue was particularly interested in them, and they ordered several dozen every week. The *Arca* survived beautifully in our tanks and aquariums, bolstering our living inventories. Some creatures would live for months in captivity, others were more perishable and had to be sold immediately.

*Penaeus* was still shaking and clattering as she dredged over the rocky outcrops. Sometimes we would go for several minutes and not hear a rattle and then she would begin to shake all over.

Then suddenly the little shrimp boat began to splutter and Leon glanced down at the oil-pressure gauge. "What the . . . ?"

Then the engine died. There was silence, frightening silence, and everyone looked puzzled. "Goddamn bitch must have airlocked or something," said Edward in a puzzled tone, "but I don't understand it. She's got plenty of fuel. It's calm, so she didn't catch air pitching and rolling. . . ."

Leon pursed his lips. He turned the key and the starter whined and whined, but nothing happened. "All right, you bitch, get your ass moving!" He turned the key again, and again the starter turned without result.

"She's airlocked," Edward said emphatically. "Sure as hell, just like she did off Shell Point. I thought I had that fixed, but that's what it is. We'd better get down there and bleed her."

The breeze from the towering dark clouds that were building on the horizon had turned into a wind. It still felt good as it gusted in my face, but I was beginning to feel uneasy. The seas were starting to rise and fall as ground swells were developing. I knew we would catch hell if we broke down out there. Leon hurried down into the engine room behind Edward and shouted up to me. "Turn the key when I tell you."

After ten minutes of tinkering he hollered, "Now!"

*Whine-whine-whine-whine.*

"Okay, shut her off."

He and Edward broke the fuel lines apart and I kept working the key. We were trying to get the air out of the fuel pump. Doug looked anxious as he poured a bucket of fresh sea water on top of the octopuses. At least the cooling winds would slow down the oxygen consumption of our animals.

"Okay."

*Whine-whine-whine-whine-whine-whine.*

Leon emerged from the engine with greasy hands. "That storm is coming up on us sure as hell. We'd better get that drag up."

He put the winch into gear and began turning the starter key. *Whine-whine-whine.* But as the starter turned the flywheel, the winch turned and a strand at a time of cable came up. Fortunately we had a big new battery down there with plenty of juice. Off in the distance, thunder rumbled and lightning split the sky and the wind gusted again.

*Whine-whine-whine.*

But the battery was starting to give sounds of fatigue, and the storm was coming closer. I picked up the radio microphone and mashed the button. "This is the *Penaeus* calling Alligator Point

Marina. . . . Please come in."

There was no answer, but I could hear voices speaking.

I repeated, "*Penaeus* to the marina, come back."

I switched channels. Thunder rumbled in the background, and I could hear voices speaking loud and clear now. It was another marina, one that was farther away, but I wanted to let someone know we were out there and having trouble.

There were two women talking. I listened for a moment. I couldn't believe it. One was actually giving the other a recipe—for butter pecan cake!

"Breaker-breaker-breaker. This is *Penaeus* calling the Shell Point Marina. We're reporting engine failure and a storm is approaching. . . ."

"And you add two eggs and a cup of butter . . ." the voice went on.

"Breaker-breaker-breaker. Would you please get off the air!" I shouted into the microphone.

"And be sure to blend it in good," the woman's voice continued. "You really have to get all the lumps out . . ."

She'd never shut up. When I cut in I couldn't block her out, we were too far away, yet she was occupying the wavelength. I waited; I listened to how to cook a butter pecan mishmash. The waves were violently rocking the boat. Thunder rumbled and lightning zigzagged across the sky. Finally she paused for a moment.

"Will you get the hell off the air!" I hurriedly interjected. "We're caught out here in a storm. Emergency!"

"Well . . . I never . . ." came back the voice. "Some people are just plain rude." And she continued talking.

I switched channels. "This is the old salty dog in Little Rock, Arkansas. . . . How you all doing down there? . . . Breaker . . . breaker . . . "

I put the microphone down in despair. I was about to give up when a voice rattled from our set: "This is Alligator Point Marina calling the *Penaeus*. Come in, Leon."

I told them our troubles.

"I can just barely hear you," the voice came back over the static. You say you're south-southeast of buoy twenty-six, about ten miles out?"

"Tell them to stand by," Leon shouted up from the engine room. "Hold it, I think we might have it. Hurry up and turn the key."

The motor rolled over with its whining sound, but it was noticeably weaker. Then suddenly it spluttered, coughed, and quit. "She almost hit . . . try her again . . ." shouted Edward.

Rain started pouring down on us and the whole sky was black, and thunder roared around us. Doug, drenched from working with the buckets and specimens, covered the buckets to keep the animals from dying of low salinity. I stayed with the engine.

The battery was almost dead, but on its last turn the engine caught. She spluttered and started clattering. A puff of black smoke blew out of the exhaust and the winch finished winding up the dredge. We hurried to get it up, but no sooner did Leon hoist it up with the cathead than we had a real problem on our hands. Loaded with rock and coral and made of steel, the dredge now must have weighed more than five hundred pounds. It began to swing with the rocking boat and we knew that if it ever gained enough momentum it would tear the rigging down.

"Get out of the way!" he shouted, and we all dove for cover. The dredge made a pass and crashed into one of the buckets of rock and sent it spilling over the deck. Leon released the rope from the cathead and the dredge crashed down to the deck, denting in some of the lumber.

It lay there in a big heap. We looked at it dumbly for a moment, holding on to anything we could while the seas tossed the boat around and Leon tried to race toward the shore. "Race" isn't a very good description for *Penaeus's* top speed, but it was doing something above its usual crawl. The blinding rain, the dark sky, and the whitecapping sea were all around us. Only the compass told us where we were heading: north-due-north, as fast as we could. The wind was now icy cold on our wet bodies and we clustered around the cabin. That was another problem *Penaeus* had—plenty of sleeping room down below next to the oily diesel engine, but very little standing space in the cabin.

The boat pitched and rolled and rolled again, and rain beat down even harder. Then we moved out of the full gale and rain for a moment, only to see ominous clouds ahead of us. "Look the way those clouds are moving up there. See how those fingers point down . . . ? That bitch might turn into a waterspout before you know it."

"What do you do if it does?" Doug asked, his teeth chattering.

"Just hold on and pray," Edward shouted above the rain. "That's what happened to Charlie Hartsfield and them on the *Isabel* a few years back. One of them waterspouts touched down and sucked all the crab boxes off their boat and ripped up everything that wasn't nailed down."

"How'd they stay?" I shouted above the wind.

"Same way we might have to," Leon replied, gripping the wheel. "Tie ourselves down. But the Isabel was a whole lot heavier than *Penaeus* is; she had a lot more power. I don't know what would happen on this boat."

"I hope we don't find out, neither," said Edward, "but you better quit running her so hard, Leon. She's heating up."

"That ain't too bad," said Leon. "She can take it, and more." But he broke the throttle down a shade.

I clutched the mast and held on; already I was beginning to feel shaky and sick. I had never been sick on *Penaeus* before, but she was really pitching and rolling. Then suddenly we had moved out of the storm, the skies ahead of us were clear, and the water began to calm down. We kept running toward land, and soon the sun was shining down again; it felt warm and good.

We undid the rope that kept the bottom of the dredge bag closed, and the four of us lifted the frame up. Carefully we hooked the block and tackle to the dredge and, straining, lifted it a few feet into the air, then shook down the rocks and hurriedly culled the specimens. Most of the crabs didn't look healthy, and several of the octopuses looked sickly from the rain beating down on them. I suspected they wouldn't survive. We hurriedly placed them in fresh salt water, hoping they would revive, and spent the rest of the trip caring for our specimens, changing water more than ever to see that any extra fresh water was eliminated from the buckets. An hour later most of our drenched clothing had dried out and, while the seas were still choppy and running with the rising afternoon tide, they weren't too bad. In another hour we'd be home, maybe an hour and a half, and that was always a good feeling, to come back and tie up to our very own dock.

It was wonderful having a dock. After a long hard day of culling specimens and getting drenched in a storm, changing water and getting baked in the sun, there was nothing prettier than the sight of the channel markers off Dickerson Bay and our dock looming up in the distance. *Penaeus* chugged along at a snail's pace, but as Leon was often fond of saying, especially to my wife, who didn't hold the boat in high regard, "She's slow all right, but when she finally gets to where she's going, she gets the job done."

We turned into the channel, and as *Penaeus* moved steadily ahead the dock loomed larger and larger, first looking like a small toothpick sticking out from shore, then light posts, pilings, and decking. There were people fishing on the dock, and our truck was in the parking lot. Even with all the care, the long trip was hard on the specimens. Some of the octopuses had died, as had a number of the rock shrimp and mantis shrimp.

Leon turned the wheel tightly and *Penaeus* made a wide sweep and started heading into the dock. "Love the way that boat turns now," said Leon. "If we could ever get her damn motor straightened out, she'd be the finest little shrimper on the coast."

The shrimper seemed to respond to his compliment, as she sidled up alongside the dock and Doug and Edward got out and tied her up. Leon switched off the engine, hurried down below to disconnect the battery cables so that the battery wouldn't bleed down from the faulty wiring system that we had never been able to fix, and we started unloading the boat.

As we lifted off the barrels and Styrofoam boxes filled with animals and piled them up into the dock cart, rain started to fall. It was a typical summer squall. The silvered weathered dock planks turned dark brown as they soaked up the rain. As we wheeled the heavy cart down the dock, the water beneath our feet grew shallower and shallower, and soon we were walking above the exposed mud flats where herds of pink and green and gray fiddler crabs were feeding and bubbling, oblivious to the light rain; too much rain and they would scatter to their burrows.

We continued. The parking lot, the truck, and the woods loomed larger, and then we were out over the marshes and oyster bars, and finally stepped out to the parking lot. Terra firma, solid ground. It's a great feeling when you've been at sea for a day, and even greater when you weren't entirely sure you were going to get back at all.

# 3

# The Dock at Night

Night is the time of awakening in the sea, the time when all little creatures come out of their burrows or crawl out from under their blankets of sand and mud to dance in the water. Night time is when the light-shy brittlestars come writhing out of their hidings and cover the sea bottoms, spreading their arms and waving them gently back and forth. Pink shrimp kick their legs and beat their pleopods and rise from the cushioning mud by the thousands to join the traffic of the night.

It is a restless period when animals are on the move. A sluggish batfish swims rapidly over the mud flat on its urgent business, the toadfish makes its high-pitched love call from its burrow, and silversides leave the marshes to swim in the open and feed upon the plankton blooms in the sea. The water is filled with fiery flashes of luminescence as tiny bits of plankton glow like blue fire when mullet or mackerel streak through the waters. During the summer the waters are hot and tepid and still by day, but they are alive and filled with creatures at night.

Anne and I were slowly carrying a large bulky wire trap down the dock. It didn't weigh much because it was made of hardware cloth and Styrofoam, but in the darkness its corners seemed to catch on every possible obstruction.

"You don't really think this thing's going to work, do you?" she asked, trying to dislodge it from under the railing.

"Why not? You can catch crabs and lobsters in traps, and they're supposed to be more highly evolved than squid. Why not a squid trap?"

"Because squid are more intelligent than crabs. The brain of a cephalopod is more developed; they're not going to swim into this thing. If you ask me, it's a waste of time and money. How much did you have to pay Carlton Messer to build it?"

I hesitated, thought about lying, and then finally owned up to it: "About fifty dollars."

"Oh, good, we've already spent four hundred dollars on a light that doesn't work, and now we just paid another fifty for a trap that won't work either. We barely have money in the checking account to pay this month's utilities. Why don't we get out of the squid business before it puts us out of business?"

"It's worth a try, at least one more. If we can trap squid we may revolutionize the industry."

We arrived at the end of the dock and interrupted a great blue heron that had been sitting on the boat stall, spearing fish with its long bill. It rose into the air, giving an indignant squawk, and beat its wings into the darkness, hoarsely crying out its displeasure at human intruders. With the heron deposed from its perch, we began setting up the squid trap.

We lowered the bulky contraption into the water and switched on the light bulb that was suspended from the stationary dock by an electric cord, then maneuvered the large wire cage so that the glow of the light bulb would shine down squarely in the center of the trap and lashed the frame to the floating boat stall with ropes.

"Bad weather would tear it up this way," I explained, "but this is just a prototype. If it works, we'll build a good one."

When the trap was finally in place, the Styrofoam rim was bobbing just above the surface and the funnels were completely under the water. We made ourselves comfortable on one of the floating docks, preparing to watch the nightly show of life.

Sitting on the floating dock in the warm evening was pleasant. There was just enough of a breeze that night to keep the gnats from eating us alive and there weren't any mosquitoes. Overhead the sky was beautiful, clear and filled with twinkling stars. We had the vastness of space above us, and the shallow sea below.

I had no sooner flicked on the light over the trap when a silverside, *Menidia*, swam to the trap, nudged the wire, and turned, leaving an empty watery void behind. I wasn't worried. It took time for creatures to come to the light. Soon the sea would be filled with life.

Downshore, against the black marshes, men and boys were carrying blazing gasoline lanterns, wading or poling small boats as they hunted for flounder. They followed the shoreline of the entire bay, illuminating the water in a most eerie fashion. Now and then we could see the silhouette of a body, or hear a distant voice calling or the knock of an oar against the side of a boat.

"They make the whole night seem enchanted," Anne said as the ghostly globes passed along the shore. "They really are beautiful."

There is more than beauty in floundering. It is a tradition that has grown over the years, a tradition of fishermen walking over miles and miles of tide flats in search of the elusive flatfish, which they catch with "gigs," or spears. It is one of the most primitive forms of fishing, perhaps dating back to man's first use of tools. Maybe floundering appeals to some atavistic memory in us that makes us feel the urge to stalk the flats as predator.

When the elliptical shape is outlined only slightly against the sandy bottom, food is in sight. A quick savage thrust of the spear, and the sand churns up from the frantic flouncing and beating of the fish. You can feel its fight, its strength and power, as the steel stake stabs through its flesh and the fish strives desperately to get away.

Unless you press your weight on the gig and pin the flounder solidly to the bottom, it may get away. But then you reach down, and your fingers feel its squirming, flat smooth body. You feel for its gills; your fingers touch its mouth, you feel the short sharp teeth and its gill rakers, and then you grab it, jerk it out of the water, flapping, and put it on your string.

At night as I walked out on the summer tide flats I often met other people floundering. Some did it for a livelihood, others for sport because it was a wonderful way to spend a night. In the little town of Panacea, where there is nothing much to do at night except watch the blue light of the television or hang around the convenience store, floundering is about the most popular recreation.

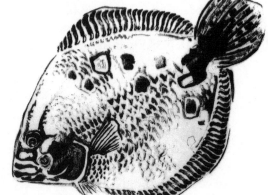

But our dock provided us with plenty of entertainment and over the years we had become more and more absorbed in the life around it. There were always things to see, whether it was pelagic creatures that drifted by the dock in the currents, such as jellyfish, or sea cucumbers that lived in the mud beneath the dock. There was life to explore on the pilings, life on the floating dock, and life even up above the dock: we could watch an osprey winging its way across the bay with a fish clenched in its talons. But at night the dock was most rewarding of all.

From the moment we flicked on the light, we heard mullet leaping and splashing in the water, and listened to the flurry of little fish and jumping shrimp. There were other sounds of night. From the water down below came the lonesome beeping call of the toad-fish. Beneath the lights we watched the tide come in and go out and always there was something different. There might be a portunid crab swimming on the surface with one claw outstretched and the other folded against its shell while it paddled along with its rear swimming legs. Their aggressiveness was fantastic: they moved with confidence and power and energy, as if daring any fish to come bother them. The next night, or even a few hours later, there would be no portunid crabs to be found anywhere.

Or the night traffic might be full of anchovies and small eels and needlefish. And then a mad whirling dervish only a few millimeters in length would come spinning up to the surface. After watching it long enough, curiosity must be satisfied: swoop it up in your dip net and you'll be surprised to see that it's only a tiny white isopod. A parasitic isopod, it normally spends its life clinging to the scales or gills of a fish, but at night it can be found swimming freely, perhaps looking for a new host. It orients to the light and thereby finds new hosts that are also attracted to the light. The isopod has sharp little hooks that it embeds in your fingers, and it holds on for dear life until you pry it loose.

Then the dance of the sea cucumbers starts, and from down in the soft mud around the dock emerges *Leptosynapta crassipatina*. This glassy little sea cucumber is so slender, so delicate, and so agile as it bends and flexes in the vertical column of water that it is hard to believe it is even distantly related to the large lumpish holothurians that live in the mud bottom and move so sluggishly. *Leptosynapta* look like a worm in shape, but they are almost transparent, and somehow strangely beautiful as they sway and dance in the light. The current sweeps by and there may be hundreds of them, undulating, changing shapes, moving back and forth in the range of your eyes, and then they are gone. If you lay your dip net down before them, the water pressure forces it open and they will glide into the distended webbing. But when the net is lifted up, you may have to look very closely to spot them. In the water the wraithlike creatures are impressive and real, but when they are hauled out of their liquid domain they are nothing but a flaccid bit of jelly.

Some years *Leptosynapta* are so common that on any given summer night, hundreds of them can be seen looping around in the current. Other years they are scarce or completely nonexistent. Perhaps they have some long-range cycle. To learn about it one would have to station oneself on the dock, year after year, recording all the various creatures that appear and disappear, and maybe after fifty years it might be possible to predict when *Leptosynapta* would have a "good year."

It seems that each year brings in an abundance of different animals that were either obscure or didn't exist the year before. One year the waters were filled with delicate pelagic nudibranchs, *Polycera hummi,* which float upside down on the surface film. I had never seen them before—they were so soft and small, and so blue in color—and I have never seen them since, although I learned that they were first reported in nearby Alligator Harbor twenty-five years earlier. There was the year of the transparent hydromedusae, *Aequorea,* and then there was the year of the acorn worms.

One summer season saw a population explosion of half beaks. They were everywhere, and even the most unobservant visitor to the dock was aware of them, with their long slender bodies and elongated needly lower jaw and cut-off upper jaw. This odd morphology enabled them to speed through water and feed upon plankton and small fish. That same season the water swarmed with anchovies that glittered like diamonds as their scales caught and reflected the light.

But I think the most impressive year of all was the year of the Atlantic threadfins, *Polydactylus octonemus.* I can't recall ever having seen threadfins in the past—but maybe I had, as an obscure little brownish gray fish with whiskerlike barbells that came up in the shrimp nets.

Then one spring they came in, not by the hundreds but by the millions. They swarmed in a huge circle at night, in a frenzied swirling movement that looked like a living whirlpool, and they kept swimming and swimming and growing in numbers until the entire sea looked as if it were going to erupt like a volcano and spew out little fish all over the earth. Sitting there on the dock, I was enthralled by this sudden appearance of strange fish. I swooped my dip net among them and caught more than a hundred. I could have collected them by the thousands, had I wanted to.

They swirled so rapidly that I felt myself being hypnotized by them, caught up in their movements, drawn along into the sea of life. When I pressed the dip net through the school, they didn't dart off in a frenzy. The school merely parted and then rejoined; that is the way of a school, how it avoids predators. A school works by confusion, offering so much food at one time that a would-be attacker doesn't know which to take first and lunges at every one and gets no one.

The threadfins bloomed, they took over; they dominated and far outnumbered every other creature in the bay. They were sucked into the intake of electrical-power generating systems and ended up in the pipes of the chemical companies down the coast, and for a while they were in every shrimp net. And then they were gone. They disappeared as suddenly and inexplicably as they had appeared, and not a single specimen has been seen since. If I didn't have a jar filled with preserved specimens to prove it, I might have thought the whole thing was a figment of my imagination. It was almost as if they had been created by spontaneous generation.

Anne was thoughtfully watching a sand eel undulating along in the water. For a moment it looked like a *Leptosynapta*, but it swam with fishlike determination right into the trap. But since it was so thin, it easily swam through the half-inch mesh wire on the other side and disappeared. Sometimes the bay was filled with leptocephalid eel larvae that were crystal clear, and only their shadows gave any hint of their presence. When we dipped them up and looked at them closely, the only visible part was a tiny pair of eyes.

There were other fish in the trap before long. Some larger minnows that swam in found themselves trapped and began banging into the steel wire. Some became gilled in the net, others just swam around aimlessly looking for a way out. After the light had been on for an hour, there were tens of thousands of tiny larval fish inside the trap that had obviously swum through the wire mesh and were swimming around and around under the bulb in a dizzying circle.

They were swimming around horizontally, the way all fish swim, but in their midst were several small golden leather jackets standing vertically on their heads. The leather jacket appeared to be another small and insignificant fish of the night traffic but it had smooth golden skin and a mouthful of sharp teeth that resembled the jaws of a mackerel. Whenever we saw this fish at night it was standing on its head, suspended in the midst of a sweeping current. When I first saw one, I thought it was sick, but when I dipped it up and put it in a bucket, it righted itself and swam horizontally, but only for a moment.

I wondered about it for some time until I met an ichthyologist who was also puzzled by this strange behavior and wrote a paper about it. By watching this fish very carefully he learned that the leather jacket made its living in part by cleaning the parasites off other fish and this strange positioning served as a signal to others that cleaning service was available.

Larger fish, like the jacks and sheepshead and croakers, stood in line while the little golden leather jacket hovered about them, biting off their parasitic copepods and cleaning out funguses from their scales with its sharp little needly teeth. No fish was observed to strike at it, no fish even swam forward aggressively to investigate it. There was something different about the leather jacket; its shiny smooth coat, its bright eyes, its odd swimming behavior—all told the tale of the services that it provided in the marine community.

It was now eleven o'clock and there still was no sign of squid. But we sat there waiting patiently, just as we had night after night. Experience had taught us that just when we were about to call it a night and go home to bed, the waters would start teeming with squid. All during the summer we had been involved with them, dragging for squid with our shrimp nets, seining for them, and night lighting.

Squid have giant axons that delight neurophysiologists, who can easily insert electrodes into the nerve bodies and study the electrical impulses that are generated. Every year these scientists, their families, and graduate students move to Massachusetts to work at the Marine Biological Laboratory in Woods Hole. The laboratory charters a large vessel that goes out and drags for squid, keeps them in large vats of running sea water, and delivers them each day to the researchers.

Squid have always been considered difficult to keep alive for any length of time, and next to impossible to ship across the country. However, one customer from Canada had contracted us to provide him with routine shipments of living squid, and we rose to the challenge.

At first, we tried dragging for them in a small shrimp net that we pulled behind the tunnel boat in front of our dock and over the mud flats at the head of Dickerson Bay. Sometimes our nets passed through schools of squid that came into the shallows to feed on grass shrimp and minnows, and we filled our buckets with them. Other times, not even one squid could be found.

But it soon became obvious that trawling was not the best way to catch squid. Even though the little estuarine dwarf squid, *Lolliguncula brevis*, was considered one of the tougher species, they were often damaged or killed by the shrimp net. We did everything we could to keep them alive, hurriedly taking them out of the net, changing water every time they jetted out a cloud of ink, and packing them into individual plastic bags and charging them with oxygen right on the boat. But even then we lost squid. The pressure of the net swooping along the bottom and the compressions of fish, crabs, sponges, and other creatures were too much for them.

So we decided to try night lighting. Scientific literature is full of accounts of squid being attracted to light. When Anne had been diving offshore on the Florida Middlegrounds at night, on the university's oceanographic vessel, she described the beauty of hundreds of glittering squid with green eyes, hovering in the beam of her underwater light, beating their tail flukes and holding a tight formation. I had seen schools of squid attracted to the lights of our shrimp boat at night, far offshore.

Night lighting from the dock was easy enough. We hung a light bulb from a rope and suspended it above the water and waited to see what would happen. At first, nothing. A few fish appeared swimming in the current and a few isopods buzzed around. And then I began to think my eyes were playing tricks on me, because there were shadowy creatures out there that seemed to have no real substance. They were like spirits down there, shadows and wraiths. I was about to declare it part of my imagination, when I suddenly saw a pair of tiny pale green eyes flash among the shadows.

I swooped my dip net down with a hard splash and when I lifted it, I thought the net had caught nothing but a mass of clear ctenophore jellyfish that had been drifting by in the current. But at that same instant, I realized that no jellyfish behaved like that, spurting water, jumping, and squirting black ink. We had succeeded. We had captured over a dozen squid, ranging from one to three inches in length.

We hurried back to the lab and put them in one of our large concrete tanks, and watched them for hours. Almost immediately they grouped together in a tight school, all moving forward or backward at one time. It didn't take them long to get acclimated to the big tank. Since they weren't damaged in the least by collecting, we didn't have a single fatality.

If squid are to be kept alive they must be fed constantly. They will not touch dead food, and they are quite selective about their live food, but no squid can resist shrimp. I scooped up a net full of grass shrimp from the aquarium and dropped them in front of the squid. In a flash of a second, almost faster than the eye can see, the school advanced forward. One squid shot out its two longest tentacles and ensnared a shrimp. Then, turning it around, holding it firmly with its sucker discs while beating a holding pattern with its tail flukes, the squid chewed the shrimp up with its powerful parrotlike beak and swallowed it. We could actually see the bits and pieces of shrimp going down its translucent body.

Our client in Canada set up large aquariums to maintain his squid and to keep them fed. We seined the marshy creeks and ditches to provide him with ample grass shrimp. However, he found that keeping his charges fed proved to be a problem. The dwarf squid were insatiable; they practically ate their weight in shrimp each day.

Between night lighting for squid, and slogging through the marsh and seining the creeks to get grass shrimp by day, we were kept busy. Too busy, in fact, and that was why we were trying out the squid trap. I yawned loudly. We had spent the last three nights out on the dock, looking at the reflection of the light bulb, and we hadn't seen the first squid. We had stayed up as long as we could, but sooner or later we got tired and went home to bed, never knowing whether or not they showed up after we had left.

I watched a small flounder undulate its flattened body up from the depths and head toward the light. It brushed against the wire trap, turned, and swam off into the darkness. Ctenophore jellyfish, *Mnemiopsis macrydi*, being swept along in the current, also crashed into the wire, and illuminated the water with their brilliant blue bioluminescent flashes, and then were swept away. I was at my wits' end. We couldn't go on sitting on the dock night after night. My last attempt at improvising a quick and easy solution to catching squid had ended in an expensive failure.

About a month earlier, after having been sucked bloodless by mosquitoes and bitten to insanity by the annoying little sand gnats that swarmed in the air when the wind quit blowing and still not catching the first squid after two sleepless nights, it had occurred to me that there must be a better way. I was getting desperate. I was bumbling around in the mornings, too sleepy to do my work. Our customer, who was in the midst of his research, was desperate for more animals. I looked at the seventy-five-watt light bulb dangling from the dock over the water, and decided that if this little bit of light managed to sporadically catch squid, a great big floodlight should do an even better job. I would install a super light on the dock. Who knew, maybe it would attract all sorts of animals. There was only one way to find out. But the light I wanted—the type that illuminates football stadiums—was very expensive. I called our electrician, who estimated that it would run up into hundreds of dollars. Yet our customer was prepared to buy several thousand dollars' worth of squid. So I looked in our diminishing checking account, fretted, and decided to defer paying even the most essential bills so that we could go ahead with the lights. Always act positively.

The next day there was a big procession on my dock—men with spools of wire and a large pole and this huge light. It was 220 volts, with 1,250 watts of mercury vapor light, guaranteed to light up the whole ocean. It should draw squid out of the Gulf of Mexico—who knew, maybe up from the Dry Tortugas! As they worked on tying it down and hooking up guy wires, they talked about its brightness. With a flick of a switch the blackest night could be turned into bright day. All day long I waited impatiently for nightfall. We were itching with anticipation.

At last darkness came, at the end of a very long day. Anne, Leon, Doug, Mary Ellen, Edward, and I walked down to the end of the dock to christen it. We looked up at the big metallic light shield that was reflecting the moonlight. The moment of truth was about to happen. With a flick of the switch the sun came on. The floating boat stalls were illuminated and the light burned down into the water. Fish began jumping and leaping out in the bay—veritable showers of them; and then there was silence and stillness.

We all waited, looking down into the water, but no fish showed up. For a moment we had seen fish around the dock, when the light came on so bright that it illuminated far down into the water. It gave more light than we had ever seen, even on the brightest, sunniest days. But then all the silvery bodies disappeared into the darkness. And there, beneath this bright burning globe that cost four hundred dollars, was a biological desert. There was empty water . . . not a living thing in it.

The fish didn't come up from the Dry Tortugas, and we didn't draw in whales and deep-sea squid from far out in the Gulf, and we certainly didn't draw in squid either. Everyone shook his head and I switched off the sun and darkness returned. Then fish started jumping and splashing all around, as if overjoyed to be rid of the unnatural brightness. Grimly I plugged in my little seventy-five-watt bulb, hung it over the water, and in a few minutes fish started to appear. Four hours later they were followed by a school of squid.

We dipped up all the squid we could, but luck wasn't running with us. The next morning we shipped them off, but instead of transferring in Atlanta they were shipped on to Chicago and delayed, and arrived stinking in Toronto.

Several days later I was sitting on the dock at three in the morning, trying to replace the order, and through my daze came an inspiration. Why not build a squid trap?

I had stayed up the rest of the night, sketching it out, designing it so the squid could swim through the four-foot-long flat funnels that would sit just below the surface. It would have to be big and deep, but it just might work.

It was now past midnight, and I was getting stiff sitting on the dock. The cool night air was seeping down into my bones and my skin felt cool and clammy. Inside the trap were thousands of tiny larval fish whirling in a mad dizzying circle, and small fish and blue crabs swam into the funnels, but still no squid. Perhaps this too was a failure, I thought. If it was, I didn't know what to do. I was on the verge of calling it off for the night, when Anne said suddenly, "Look . . . next to the piling . . . over there . . . squid, a bunch of them!"

Sure enough, just outside the illuminated water, I could see the flattened shadowy forms just below the surface. I could catch only the slightest glint of their green eyes. I couldn't tell how many there were, but there was certainly a school of squid about six feet away from the trap. They were hovering on the surface, beating their tail flukes and holding a tight formation as they watched the tiny minnows circling the center of the trap.

I reached for my dip net.

"Don't disturb them," Anne cautioned. "Let's just watch them for a while and see how they react to the trap."

"All right, but if they don't go into it pretty soon, I'm going to swoop them up," I said nervously. "We have to fill that order or we'll be in trouble. . . . Come on, squid, swim through the funnels. We have to pay for that dumb light!"

Then suddenly one of the squid raced forward and swam right through the funnel, and hovered there among the swirling fish. A second later I saw the shadowy forms of ten more squid bunched together in the very center of the trap.

"They'll swim out, I'll bet they will," said Anne. "No squid is going to be trapped like a mere blue crab."

But they didn't. They instantly realized that they were surrounded and they followed their normal instinct. They sounded to the bottom until they hit the wire floor. And there they stayed until we scooped them out with the dip net.

"Phooey," said Anne disgustedly. "I'm glad we figured out how to catch them. But who would have thought it. Squid sure went down in my estimation!"

I was afraid they might go into a frenzy and beat themselves against the wire and we would have damaged specimens that weren't fit to ship. But that wasn't the case.

Not a single squid managed to swim back out of the entrance funnels after they came in at night. However, we learned that if we didn't get them out of the trap immediately after the sun rose, they did manage to get their bearings and leave as a school. It was only the light burning down through the trap at night that kept them confined.

So we could now leave the light burning on the dock all night and catch squid while we slept. The only trouble was that during certain seasons there were no squid in the bays, and if the tides and the winds were wrong they were scarce, but both we and our squid customer had to accept that.

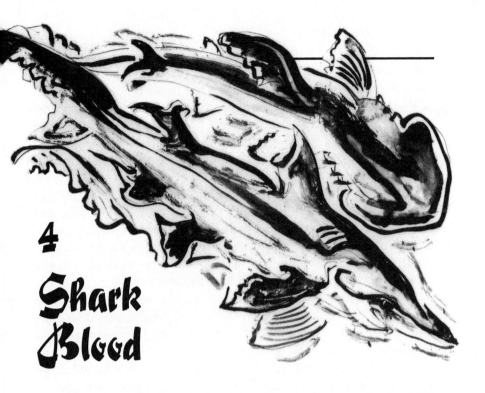

# 4
# Shark
# Blood

The letter from the University of Wisconsin's medical school was emphatic. If we didn't ship the shark blood by September 8, we had better forget it. And four hundred dollars for a gallon of blood was good and much-needed money. We had tried shark fishing over the past few weeks and I was getting more and more discouraged because we were catching small sharks of mixed species, which did not fill the requirements of the order. When I figured up all our time and expenses, I was beginning to wonder if we would break even if and when we succeeded. I tried to combine shark fishing with dredging, or leaving jugs off a sandbar while we were tide-flat collecting, but it never worked.

Over the years I have found sharks to be a losing or at best a break-even proposition, although I could never turn down an order for one because of the thrill involved in catching and handling them. Never will I forget the order for a pound of shark brains; I sat there ripping the frozen heads of black-tip sharks and scooping out the tiny amount of soft matter. It took more than ten sharks of assorted sizes to fill the order. Then there was the order for twenty pounds of shark liver, which would have been easy to fill and highly profitable in July or August, but the order came in February, when no sharks were to be found anywhere. Nevertheless, we spent hours fishing for them in deep water, hoping a big stray would be around. We lost the order.

Once in a while we won one, though. The National Institute of Health wanted a hundred pounds of hammerhead shark skin and fins, and we happened to meet a fisherman who had just entangled a fifteen-footer in his net. The shark ripped and twisted the three-hundred-yard gill net, but finally after three hours died, and the fisherman triumphantly beached it not more than a mile from my dock. That too was in August, which is a very good time to catch sharks. I squatted down in the boiling sun, sawing away and skinning the monster while a mob of people gathered round to ask what I was doing and why. By the time all the fins and the two elongated sides of the hammerhead's hammer were sawed off, and large sections of skin were piled up, I was sweaty, reeking of shark, and exhausted, but I had my hundred pounds of skin and could have saved another hundred if I'd wanted it.

But whatever profits were gained by selling three hundred dollars' worth of hammerhead to N.I.H. were probably lost when we went out for forty pounds of shark cartilage. Who would have thought it would take four six-foot sharks to fill an order like that? A shark is supposed to be all cartilage and muscle, only that cartilage is less than five percent of its total weight. Sometimes you can fish all day and not catch a single shark, and the next time the water is festering with them. It takes time and labor, tank after tank of gasoline, rigging lines, fresh bait, and your undivided attention, and still there is no guarantee that you'll get the right shark.

So why didn't I write back to the University of Wisconsin and say that we could no longer afford to ship sharks or shark products? I suppose it was because I am enthralled by them. I love their sleek form and beauty, the way they move majestically through the water.

Now and then we would bring a shark back and put it in our tank, and there was no fish that could compare with it. So little is known about sharks even though there are a number of laboratories working on their behavior and physiology. There is an innate fascination about them, a respect and fear at the same time. And in this day and age, when man with his bulldozers and his power and might is pushing back the forests, draining the swamps, and killing off every other species, it is nice to think that there may be one species that he will never master. No matter how far technology advances, there will probably be sharks out in the ocean, roaming around in the depths, free and untamed. The big crocodiles of the African rivers will have all been killed off, the lions and tigers will have been shoved to the verge of extinction, with only a few remnants in zoos and national parks, but the sharks will still roam the deep unfathomable oceans.

Then there are times that I have been rather blasé about sharks. I have taken tiger sharks, lemon sharks, bull sharks, hammerheads, and black-tips from the estuaries and waters of Florida. I have netted small sharks in the gill nets and have on rare occasions successfully supplied them alive to large public aquariums. It amazes me how fragile sharks really are, how easily life slips from their bodies, especially when you want one to live.

I recall the many hours I have spent walking freshly caught sharks around our tanks, moving them so that the oxygen and water can pass through their gills, and then at last there is that spasmodic jerk, a swish of the fin, and the shark begins to swim . It swims a few feet and then sinks to the bottom and we start walking it again, and again until finally the shark is able to swim under its own power and moves freely around the tank. But in all likelihood when we would come in the next morning, it would be lying belly up on the bottom of the tank, stiff and dead.

It is almost as if there is a free, uncaptive power about sharks, as if they aren't meant to be held in captivity and beheld by man in a cage. They are a power of life and movement, and when confined that life slips away from them. They are so quick, so sensitive and so alive. Even their skin, leathery as it is, with its sandpaper texture, is alive with tiny electric organs that give off a low-level voltage which enables the shark to know where it is and to locate its prey in murky waters. It has no bone, just cartilage or gristle, and if you handle it roughly at all, you can cause internal hemorrhaging and dislocate its soft tissues.

On that hot August day the sea was mirror calm and we could see for miles across the hazy horizon as we raced across the surface, skimming it, shattering it with our wake. We had the tunnel boat all prepared for shark fishing. Not only was it filled with bloody bait, hooks, lines, anchors and jugs, we had Styrofoam chests filled with ice and gallon jars with tiny bottles of heparin, which would be used to keep the blood from clotting. That was very important because shark blood clots easily. Our customer insisted that the blood cells be evenly distributed throughout the plasma, not one big ugly clump of cells.

We didn't have to run very far before we arrived in excellent shark-fishing territory. It could even have been called "shark-infested waters," but that definition I find a little vague, since any body of sea water on this planet except the Antarctic waters is "shark-infested." Along the far end of the Panacea Channel, where the water was deeper and the current was swifter, the largest number of sharks could be chummed up in the shortest possible time.

Leon slowed the tunnel boat and we glided to a stop and he began to chum. Beneath our boat was a magnificent sea-whip bottom, a limestone outcrop of flat rock that was so heavily overgrown with sponges and gorgonians that it looked like a dense forest of yellow, purple, red, and orange plastic-coated wire popping up from the bottom. It was scarcely fifteen feet deep, and we often went diving there to fill our orders. We had from time to time also pulled some huge sharks from these waters, including tigers, hammerheads, and bulls.

But never while diving had we seen even one shark. In fact, in all our combined years of diving, our shark sightings had been extremely rare. I recall having met a young hammerhead once while diving around some pilings. I was absorbed in pulling sea pork off and when I looked up I saw the hammerhead approaching. When it came within range, it became startled, and quickly turned and fled. Anne's encounter some years later was a little more impressive. She had been diving around an oil rig off the Louisiana coast, making a survey of fouling organisms, and a small sand shark, about four feet long, appeared, circled, and refused to go away. He appeared territorial in his manner, and seemed to regard the large funny-looking divers with their clouds of bubbles as trespassers. When he showed a little too much interest, Anne and her buddy gave a thumbs-up signal and surfaced. All her other dives, before and since, have been sharkless.

But we knew one sports diver who frequently dove along the north Florida rocky outcrops and he always saw sharks, sometimes enormous ones, and he had some close calls. He was a spear fisherman. He went down to the bottom not to observe and appreciate but to kill. He would swim up to a large friendly grouper that peered out at him curiously from its cave, and send his spear smashing into its side. The frantic death struggle of the fish, and the cloud of blood permeating the water, invariably brought sharks to the area. Once this mighty he-man had to kill a ten-foot bull that came a little too close.

Perhaps it may appear a bit hypocritical for me to write about my adventures in catching and killing sharks, and then berate someone for doing the same, but there is a difference. When a man straps on his scuba tank and jumps overboard, he leaves his own world and ventures into another. He should behave as a guest of that world, take only what he needs, and destroy as little as possible. When I see so-called sportsmen reeling in a struggling shark, shooting it, and then cutting it loose and letting it sink limply to the bottom, I feel there is some kind of crime against nature being committed. If the shark were used somehow—for food or for science—then somehow taking away its life wouldn't be as bad and wasteful.

The so-called primitive people around the world who live with their environment have a veneration for the creatures they catch and eat, whether it is a sea turtle, a whale, or a shark. Often the catching and slaughtering of a large animal is accompanied by ritual songs and dances of appreciation to the various sea gods and nature spirits that have provided them with food and medicine.

So, as I sat upon the stern of the tunnel boat, ramming large steel-barbed hooks through chunks of bloody fish, I was asking the sea to give up one of her sharks. And in return, I would do everything in my power to keep my fellow human beings from destroying and polluting the sea.

"We ought to catch a hell of a shark here, with all this chum," Leon said as his knife sliced through the bloody red meat of a crevalle jack. "I used to have a line of crab traps here, and we'd see some he-mongus monsters following the boat. I saw a hammerhead here once that must have been fifteen feet long. If we caught one like that we'd have enough blood to fill a washtub. All right, Edward, start throwing out the chum."

That morning we had made the rounds of all the fish houses and saved two barrels of their discards. There was a great array of bloody bait including mullet heads, backbones, guts and scales, and half a garbage can of semirotten shrimp heads. But for bait we bought bloody jacks and fresh mullet, nice sleek blue and black fish with firm tissues, packed on ice. They were good enough to eat, in fact, so I made sure that they were well iced just in case we didn't catch any sharks, or caught sharks before we used them up. It had been a while since I'd had some nice fresh mullet for dinner.

Long ago I had learned that it didn't pay to be stingy and get bad bait when shark fishing. Although there would be a few sharks that were chummed up and out of their minds in a gobbling fit, most sharks are rather discriminating. If we economized and put old fish on the hook, it would sit there until the crabs ate it. We would see sharks circling all around it, but seldom would they actually strike. Only a fresh firm healthy fish, the type you would eat yourself, could really guarantee a shark. The only thing better would be to put on a live fish and let it thrash around; its distress movements and vibrations would bring in sharks before you knew it.

I have to admit that I was a little grudging as I took those handsome mullet out of the ice chest and handed them to Leon, who rammed the hooks through their bodies and tossed the jugs overboard and then threw out their attached heavy concrete blocks. Mullet was one of my favorite foods. Edward dumped a small bucket of mullet and shrimp heads into the sea and the bloody juices clouded the water. "If anything's going to draw sharks, this will," said Leon excitedly. "The tide is about right too. It's still on the rise and that's when sharks are feeding."

He threw out the second jug, raced several hundred feet ahead and threw out the next jug and the next, until they were all sitting on the flat calm surface in a neat row. Around each jug was ten pounds of chum filtering through the water, dispersing and spreading and hopefully reaching the nostrils of large hungry sharks. Sharks' powers of smell and taste are amazing. When there is blood in the water they will come from miles around. But how do they do it? I wondered. How does a little bit of blood that mixes with millions of gallons of water and spreads out to only a few tiny molecules still manage to stimulate the olfactory nerve of a shark and send it zeroing in on the kill?

My thoughts were interrupted when the first jug we put out began to pull and bounce and Edward cried, "We've got one! We've got one!" Leon jerked the starter cord. The outboard did nothing, and he cursed and jerked it again, and this time it fired up. He hurriedly slapped it into gear and we were racing off toward the jug. While it was jerking and thrashing about in a wide circle, it was obvious that whatever shark had struck, it wasn't big enough to get up and run with the jug. It could barely pull it under, but that didn't stop it from moving it around so fast that it was difficult to grab.

Each time we reached out to snatch the jug, it pulled down and moved off, and then suddenly it gave a hard snatch and went way down. We looked at the jug, a bit puzzled; it didn't make any sense. Then fifty feet away from the boat the jug surfaced again, but it was hardly moving. All the fight had gone out of it, yet, judging from the way the jug sat heavily in the water, there was still a shark on it.

There was a shark on it all right, or at least a piece of one. A small black-tip, about four feet long, had taken the bait and been chopped in half by a much, much bigger shark. Somewhere down there was the shark we wanted, one capable of taking a shark that must have weighed forty pounds and severing it with a single bite.

"If we can get the one that bit this one," observed Leon thoughtfully, "I think you'll have all the blood you need."

Edward jerked the hook out of the dead shark's mouth and hooked on a new mullet and tossed it overboard. Now we were watching the jugs intently, our eyes roving from one jug to another. Nothing moved. Somewhere beneath our boat lurked a big shark; perhaps it was a huge hammerhead or a tiger. Still the jugs didn't move. Perhaps our big friend down there was only interested in frantic struggling game. Perhaps dead fish didn't appeal to its gourmet taste. We threw the badly bitten shark overboard and as I watched it sink to the bottom I hoped that I might see a large gray shape appear and gobble it down. But the dead shark sank out of sight.

Although I had a face mask in the bow, I wasn't about to put it on and stick my head into the water. The smell of shark, the feel of shark, made me tingle. Perhaps it aroused some atavistic fear in me, dating back to a time when ancient people lived at the seashore and made their living spearing fish, catching lobsters, and digging clams. Food would have been plentiful at the edge of the sea, but their greatest terror would have been a dorsal fin cleaving the water. I thought of those tremendously powerful toothy jaws down below, clamping down on the rest of the dead shark, and I knew that I didn't want to fall overboard.

The chances of getting attacked by a shark are about as great as getting hit by lightning. Yet in reading the popular literature one gets the impression that sharks have nothing better to do with their time than rove the sea looking for humans to eat. Every day people are mutilated on the highways in the machines of their own making, ripped apart in airplane crashes and burned in fires, and slaughter each other, and it is all taken for granted. In fact, when you read that a hundred people were killed in a train crash, the news is regarded with a blasé "tisk-tisk, what a shame," and you go on about your business.

But should one large predatory white shark or mako choose to partake of a human, it becomes headlines all over the world. The thought of being eaten, with large sharp white teeth tearing away your flesh and crunching your bones, is so horrible that even the most experienced divers shake at the thought of it. The word *shark* will send people stampeding out of the water in terror. Yet these incidents are rare; only a handful occur each year.

Actually, even the great white shark, often called "man-eater," spends most of its time eating fish and squid, with an occasional sampling of sea turtle or sea lion. Sharks are very much a part of the environment, they are not horrid monsters from outer space sent out to destroy the world. Although shark ecology is a long way from being understood, sharks provide a much-needed service in the sea. To some degree they are cannibalistic, particularly when one of their kind is wounded or caught in a net. Little sharks are often found in the guts of bigger sharks.

If a large hammerhead is cut open, its gut will often be full of stingrays and its mouth perforated with the rays' venomous stingers. In all my years of working in the north Florida waters and traveling around the world, I have never met anyone who had been attacked by a shark. I have, however, carried a number of screaming miserable people to the hospital with stingray wounds in their hands or feet. Every year along the Gulf and Atlantic coasts there are hundreds of stingray wounds inflicted, and there is no estimating the number of catfish wounds, which are so painful that one practically faints when perforated by them. People walk along the shallows and step on stingrays that are buried in the sand, and they are wounded by catfish when they take them off their lines.

Sharks love catfish. Probably they're the only thing in the ocean that can consume them with impunity. There was once a shark-fishing tournament in Apalachicola, and large numbers of sharks were brought in and cut open by people ghoulishly looking for human remains, but they didn't find any. What they did find in the guts of twelve-foot tiger sharks and ten-foot lemons were hundreds of partially digested spiny catfish. Every angler and every shrimper I have ever met, myself included, can tell stories of excruciating wounds from catfish. Where man dumps garbage into the sea, the waters are often black with vast schools of catfish.

Nevertheless, there is still that fear of being carried away in the jaws of a big shark. Me, I am *human*, I am the most important thing on this planet. The indignity of it all! But then, man seems to despise predators in general. Systematically he has hunted down the mountain lion, poisoned the coyote, shot eagles, and practically exterminated the big cats of Asia and Africa. The crocodiles that infested the tropical rivers are now practically gone. Snakes are shot on sight. It isn't a rational reaction, it's an emotional one.

One reason might be that man is a top-chain predator, and he therefore despises all other competition. A rattlesnake eating a cottontail rabbit must be shot because the snake is potentially eating *his* rabbit. But considering that there are no vicious wars between wolves and panthers, and that other top-chain predators seem to live in harmony, I must put that theory aside.

Perhaps our atavistic memories have survived over the hundreds of millions of years from the time when we were tiny helpless creatures evolving in a sea of large fish with big nasty teeth. Down through the millennia, through the vast expanses of time, we have been eaten again and again and again.

My thoughts were interrupted when suddenly a large black fin popped up a few feet away from the boat and I shouted, "Shark!"

"That ain't no shark. That's a porpoise," said Edward, laughing. I felt foolish. Sure enough, the fin came up and the black back rolled and I could plainly see that it was a porpoise feeding around the bottom. Very likely it was eating some of the mullet heads that were dumped down there, because bottlenosed dolphins, as they are properly called, are very fond of mullet. Another porpoise surfaced and blew a loud *phoof* and went below.

We sat there drifting for almost an hour. Once we went over and checked the bait and threw out more chum, but there was no more activity. It was beginning to look like another wasted day; obviously the tide changed, and it appeared the sharks had headed out to sea.

Then suddenly Edward shouted, "We got one! We got one!" and pointed his finger out toward the horizon at our farthest jug. "Look at that son of a bitch fight!"

The slick calm waters were splashing and beating, and the big white jug shot down under the water and in an instant came exploding to the surface. Even from that distance we could see the big gray tail fins lash out of the water. "Let's get him, let's get him before he gets away," cried Leon as he jerked the starter cork. The motor responded with a loud roar. He turned down the throttle enough to jam in the gears and we were off whizzing toward the rapidly moving jug, gaining on it, overtaking it.

The tunnel boat was the best kind for shark fishing, since its motor was mounted in the bow, not on the stern. Leon could get right up on top of the jug no matter how fast the shark swam, dragging its concrete block behind it. If the shark veered off when it heard the motor approaching, Leon could turn equally fast. "Jack, get ready to shoot him, you may not have but one chance."

I stood there waiting tensely with my arm outstretched and my pistol pointing at the water. Leon reached down and grabbed the jug after it surfaced and braced himself. The shark sounded, and it almost wrenched him out of the boat, but he held on and yanked backward. For an instant I saw a gray spotted form appear beside the boat and I squeezed the trigger. Down went the shark, this time jerking the jug out of Leon's hands.

"All right, all right," he said irritably, "I'll get you this time. Jack, wait until I get him to break water this time, then let him have it."

Leon stepped back into the stern where Edward and I were able to grab the jug with him. "Now don't let that rope wrap around your hand," he warned us. "That son of a bitch will pull you overboard. If we can't hold him, let him go."

They pulled with all their might, and this time I could plainly see the shark break water. It was a tiger shark. I saw its great broad head, the streamlined ridges of its caudal peduncle and tail, and the distinctive grayish white blotches over its silvery body. There was something magnificent about it, something exciting and wonderful in the way it was so efficiently constructed and so powerful. When they hauled back with all their might I saw its head, its jaws with teeth protruding in all directions, and I squeezed the trigger.

I felt the .22 explode in my hand, and at the same moment I saw a little red splotch appear in the shark and heard the "splat" of the bullet as it smashed into its flesh. The shark went wild, it sounded, and Leon and Edward let go. The jug was bobbing around in a dizzying circle, but the tired wounded shark didn't have the strength to run any farther, and I knew that now it would only be a matter of time before we had him. We waited for a moment then closed in and lifted the jug, but there was more fight left in the shark than we had anticipated.

When we pulled him up a second time, I fired again and again, putting three holes in its head. But I knew I had botched it, because in my excitement I had fired too low and the last bullet had gone into its heart. A great cloud of blood began to spread around the boat, pouring out of the wounded pain-crazed animal.

This time the shark didn't break away. It struggled but it was dying. As we fought against the eight-foot monster, hauling and pulling it, I felt slightly sick at the blood lust within me. Each time the shark fought less and less, and finally when we all pulled back and raised its head out of the water, I pressed my pistol up against its skull and blasted out its brain. I wanted to kill it as quickly as possible. I wanted the life to leave the shark now, so that it would be dead meat and not a living thing any more.

As I helped haul it out of the water I didn't feel like a hero. I knew that I had killed a noble creature that had roamed the seas for God knows how many years. No one knows how long a shark can live. Where this shark had been no one would know either. It was a worldwide species. It may have wintered in South America, or it may have come from the Bahamas. For all I knew it could have come from the Indian Ocean but now its life was ending and by my hand.

The shark was a long way from being finished, however. Even with its brain destroyed, muscles and nerves were still functioning. For a delicate, easily damaged animal that probably wouldn't have survived a week in an aquarium without the most intensive care, this tiger shark put up quite a struggle. While Leon pulled the head out of the water, Edward and I grabbed the pectoral fins and hauled backward. The shark was slapping its tail wildly and gnashing its jaws. It was obvious that the three of us could now swing it into the boat.

"Hold this," said Leon, passing me the rope that was attached to the hook in the shark's mouth. While Edward and I struggled to keep the shark pulled half out of the water, Leon tied the anchor rope behind the two pectoral fins. The straining and hauling were making the shark bleed more profusely and I looked at the red water around the boat, knowing that the precious resource I was seeking was wasting away into the sea. Yet there appeared to be plenty of blood. If we could only get him up, then we might be finished with this sanguinary business.

With the shark tied securely, we all grabbed the fins and started hauling it backward, straining with all our might to pull it out of the sea. "Now watch that son of a bitch," warned Leon. "He'll bite your leg off before you know it. When we get him in, stand clear."

We gave one mighty pull, with all our combined weight counterbalancing the shark's, and it emerged from the sea. As it came up over the side, the low gunwales of the tunnel boat pressed dangerously close to the water and for a moment some splashed in on us. "Let's not capsize this damn thing," cried Edward. The shark slapped the water with its tail and slid into the boat with us. It went wild, pounding the plywood deck with its powerful tall, it wriggled to and fro, its powerful jaws mightily snapped shut. And every time that tail slammed into the side of the tunnel boat, I could have sworn that I saw the seams getting knocked apart. The shark flounced and bounced and shook and the little boat bobbled around.

I looked down at the blood on the deck. "Come on, let's get him up somehow, we've got to bleed him before we lose it all."

"I hate to fool with that damn thing," said Edward. "Let's let him lie there for a while and finish dying."

"We can't," I said. "We have to get the blood out now, while his heart's still beating."

We tied a rope around the shark and grabbed it by the two pectoral fins, which made handy handles, and hoisted it up on the little metal boom in the stern of the boat. We used the boom to tow shrimp nets and haul drags, but it wasn't designed for such a heavy downward pull and almost bent under the strain. With all our weights to counterbalance the shark's bulk on the other end of the block, we succeeded in lifting it straight up so only its tail lay on the deck. Leon tied the shark off and it hung there limply, with only an occasional feeble twitch. Hurriedly I pulled out a gallon jar and pried off the lid of the heparin solution and dumped it in. Then Leon started sawing through the tail to get to the artery, and as soon as he did, the blood started gushing out. By this time I was thinking only of the blood and not of the shark, and I was happy to see the jug filling up with red liquid.

Suddenly the shark came to life. Although Leon was holding the severed trunk of the tail steady to keep the blood gushing into the jar, he wasn't expecting that last surge. The tail swung out and knocked the jar from my hands, and the blood spilled all over the deck. At the same time, it bashed me in the face and sent me sprawling backward on the deck. I was lying in the middle of the blood; it was all over my neck and face, and I could smell its sharp uric acid. The bruised skin around my eye began to swell. It felt like I had been hit by a sandpaper punching bag. I knew I'd have a whopper of a black eye, but what the hell, it would be a good story.

I struggled to my feet and hurriedly picked up the bloody jug, cursing as I dumped in more heparin. I jumped back on the shark. It was still bleeding, but I had lost most of its blood. Even though I squeezed its tissues and massaged it and worked it over and over, trying to push the blood out, it was clear that we wouldn't get a half gallon. I looked at all the blood over the deck, all that valuable blood, and I looked at this big creature of the open seas that had died in vain, and I felt lousy.

"We should have lashed its tail down," said Leon thoughtfully. "I sure hate that. But with the next one we'll do better."

We looked out at the jugs and we saw one moving, and raced over to it, but by the time we arrived the jug was still. We lifted it up and the hook had been snatched off. Leon tied on another big hook, one that measured a good four inches, and baited another fish. Then he went over to the next jug and pulled it up and there was only a bare hook. When the sharks were in and feeding, keeping bait on the hooks was a chronic problem. Little sharks were very adept at moving in and snatching the bait off without getting hooked.

Only two jugs still had the original fish on them, but they were now waterlogged and bloodless, so we replaced them with fresh mullet. Then in the middle of the jugs we dumped the tiger shark overboard, after slitting it open so that the other sharks would be attracted to it. I watched it sink down out of sight like a stone, and I felt myself saying farewell to it. I was so very sorry to have sacrificed it and not to have used it wisely.

I had learned a great deal of respect for tiger sharks over the years. I thought they were rather attractive creatures, even more attractive than other sharks, with their splotched silvery gray bodies, the flattened ridges that run down their sides, and their powerful muscles. In the aquarium there was no more distinctive and aggressive-looking animal than the tiger.

Tiger sharks are probably the least discriminating sharks in the ocean. They have been implicated in attacks on man, and once in Honduras I spoke to a boat captain who had witnessed an eighteen-foot tiger bite a cow in half. Frequently we found horseshoe crabs, stingrays, and nearly anything else in their stomachs. Before we sent this monster to the deep, I hurriedly slit open its stomach to see what it had been eating; after all the digestive juices and glop poured out, I saw that it had eaten six large spiny boxfish, which were now partially digested. The fish were still inflated and bristling with spines, but it hadn't done them any good. Against most other predators the defense mechanisms of a spiny boxfish are excellent. When it is attacked it rapidly gulps water and inflates like a balloon and the little spines on its body become an armament of bristling thorns and the predator lets go. But the tiger didn't mind a bit, he just wolfed them down.

From the gut we also pulled out a plastic bag and several paper sacks that it had just consumed. It was some of the paper garbage that we had dumped overboard. When we poured out the barrel of fish chum we noticed the plastic bags and garbage sinking into the depths along with the fish heads. Tiger sharks are notorious for gobbling things up, and all sorts of indigestible man-made materials have turned up in their guts.

There is only one species of tiger shark in the world, *Galeocerdo cuvieri,* and that makes them easy to identify. That's what I liked best about tiger sharks—they didn't require endless hours of trying to key them out, or hiring an ichthyologist to come down from Florida State University to make positive identification. When supplying customers with organs or blood, I had a responsibility to see that the shark was properly classified.

I knew that if we pulled up a big sand shark next and drew out its blood, we would have to haul the carcass in or call someone from Florida State University and pay him a consulting fee to come down with his textbooks and keys. The specialists would open the mouth, examine and measure the teeth, check the angle of the dorsal and caudal fins, look at the gill rakers and count them, and fret and worry over it for a long time before they came to a final determination. All sand sharks belong to the genus *Carcharhinus,* but there are a dozen or more possible species. It could be *a Carcharhinus milberti,* the sandbar shark, or it could be *Carcharhinus leucas,* the bull shark, or *Cacharbinus limbatus,* the black-tip shark.

The distance between the eyes, the shape of the dorsal and caudal fins, and serrations of the teeth all make a difference distinguishing one species from another. But even a child could identify a tiger shark.

I cursed myself for my carelessness. I was so close. If only I had managed to hold on to that shark's tail and not spilled the blood. All I had to show for it was a swollen eye and an abraded cheek.

While I was feeling sorry for myself, another shark struck and this time the jug shot way down and under. When it came up we saw it speeding over the surface. "That's a big one," yelled Leon joyfully. "From the way he's moving that jug and dragging those cinder blocks, he must be a heap bigger than the last. I'll bet anything it's another tiger."

We had a difficult time chasing the jug down. It took almost an hour before the shark tired and we could even grab hold of it. When Leon finally caught up with the jug, he pulled it up and hurriedly wrapped it around the bow post, and the shark proceeded to tow us around the Gulf. At last the monster was exhausted, and we were able to pull him up to the surface. I squeezed off a shot and hit the shark's nose. With an angry desperate splash it shot down into the depths, snatching the rope out of Leon's hands.

"He's going under the boat," Leon said. "Goddamn, he'll get tangled up in the prop." He shut the engine off, and it was a good thing he did. The outboard motor bounced on its shocks where the shark line had tangled around it. We tried pulling the motor foot up out of the well, but the shark rope was wrapped around it. "We'll just have to unwrap it," said Leon. "Give me a hand, Jack."

We squatted down alongside the well and looked down into the bluish green water, and Edward pulled the motor up as far as it would go. There we saw the nylon rope twisted around the prop. Leon reached down and grabbed the rope, then pulled the shark up to get some slack on it and handed me the rope. Then he started unraveling the rope as I held the shark and released the tension from the section he was working on. I was watching absently, looking down into the water, when suddenly I saw a large grayish silver form appear directly under the boat. It was a shark headed right for Leon. I couldn't yell. It was happening too quickly. Leon was alert. He spotted it and jumped back. We heard the jaws snap shut where his hands had been, and the shark passed under the boat and came up alongside us. It wasn't the shark we had hooked, it was another shark entirely, one that must have been ten feet long.

"Goddamn," said Leon. He was white and shaking. "I almost lost my hands. That damn thing almost bit me. I ain't never been so scared in all my life!"

"That must be the mate to the other shark," volunteered Edward. "Maybe it was trying to get the other one free."

"I doubt it. It was just a shark and it saw a pair of hands in the water and struck at them," I said, but I too was shaken. Never had I seen anything happen so fast, that big shape appearing out of nowhere.

After a while the hooked shark lost some of its fight as the bullet in its head took effect, and we were able to free it from the wheel. The three of us hauled it up and even though it was bigger than the first shark, we dispatched it with a no-nonsense attitude. As soon as it broke water, I carefully aimed my shots and blasted away, demolishing its marble-sized brain.

This time, when we cut the caudal fin off and hoisted the shark up, we lashed the tail down firmly. There was no way it could waste blood. All my bullets had gone into its head, and there was no damage to the heart and the shark had plenty of blood. Blood gushed out of the tail in a quarter-inch stream, filling the gallon jug right to the brim. I shook it and stirred it to make sure that it mixed with heparin, and I could see that it was doing fine. There was no sign of coagulation. When the jug was filled to the top, I screwed the lid down and stuffed it in ice. The shark was still bleeding profusely, so Edward handed me another jug. Why not get two gallons, there was no sense wasting it.

The second jug was almost filled to the top when the shark quit spurting and lay there twitching feebly as the last gasp of life escaped from it. As we headed back for the dock, with our Styrofoam chests filled with the jugs of blood, it occurred to me that this was really a dangerous business.

Collecting hermit crabs from the tide flats, dipping up jellyfish from our dock, or taking off barnacles from the pilings—the lesser and more humble creatures of the sea—had a lot less risk than playing vampire to a tiger shark. But there was still a lure and a challenge that sharks presented, one I couldn't resist.

I knew that when another order for sharks came in, I might grumble about it, but in the end I would probably end up filling it, or trying to.

# 5

# Death in a Salt Marsh

During the early fall, night was the most comfortable time to work, and even though our squid trap did much of the work for us, we still enjoyed sitting out on the dock at night, dipping up creatures as they drifted by in the glare of the lights. It was pleasant out there, often there was a cooling evening breeze, and we could lie on our backs and watch the stars glistening in the sky.

But sitting on the dock could be unbearable by day. There was a stagnant haziness about the air as the sun beat down relentlessly. The vegetation in the marsh around the foot of the dock had a dusty look about it, the water was tepid and hot, and the fish were scarce.

The sea became murky with plankton that kept blooming as the sun broiled down. It wasn't exactly lifeless in the water. There were schools of glittering flashing menhaden down beneath the pilings, needlefish still patrolled the surface, but there was a noticeable reduction in the diversity of life about the dock.

September can be as hot as the middle of July. As we trudged over the mud flats, filling our orders for fiddler crabs or seining the marshy creeks for killifish, we often wore big straw hats and long-sleeved shirts and trousers to keep from getting broiled to a crisp. The water glared up at us with its brightness, and even though we turned dark brown over the summer, it seemed that new sunburns were born in the early fall. The normally enjoyable art of collecting specimens became an exhausting nightmare and all we could think about was getting back to the coolness of the laboratory. If we seined a creek for grass shrimp, most of them would be dead before we could finish getting them out of the net. We learned to throw bags of ice into the bucket if we hoped to get anything home alive.

It was amazing how barren the tide flats became, incredible how difficult it was to find normally abundant plumed worms or burrowing anemones. Even moon snails and small horseshoe crabs were scarce. Everything was fleeing the dog days. But it wouldn't be long before that first hard chill of cold wind would come sweeping down from the north to get things moving. When the water temperature dropped, fish would be on the move, mullet would start to school, and trout would be foraging around the grass beds. When that first chill came, the sea would be reborn.

But there was no hint of that chill in the air. The sky was bright overhead, and there was no sign of winter clouds anywhere. Only the hot summer clouds were there, huge cumulus clouds that pyramided high and puffy into the atmosphere, and seas that were calm and glassy.

Schools started in September, and when school was in session, Gulf Specimen Company was busy. Students were arriving back on campuses, professors were planning their labs for the first semester, and researchers were starting on their year's work full of energy, fresh back from vacation. And so Leon and I were walking down the beach gathering periwinkle snails and fiddler crabs while Edward, Doug, and Anne were back at the laboratory putting specimens into plastic bags for shipment.

"I don't know what we're going to do if this hot weather keeps on," Leon said, wiping his brow. "We ain't got half of the stuff our customers want. There ain't no flatworms out there, no cloak anemones, no red-beard sponges, nothing. They think just 'cause we list it in the catalog we can just pull it off the shelf, but it don't work that way."

"I know it," I replied, as I slogged through the mud, "but we'll just do the best we can. We'll just give substitutes when we can."

"Hell, we're running out of stuff to substitute. This is about one of the hottest Septembers I can ever remember. But you can mark my words, we're going to have one hell of a bitchy cold winter."

"Think so?"

"You can bet on it. When the marsh grass goes to flowering this early and those high tide bushes turn white and the fish crows start bunching, there's going to be a bad winter. And that ain't the worst, old man Spivey told me he was crabbing down at the lighthouse and he saw all kinds of blue crabs swimming on top of the water. That means bad weather's coming for sure."

"Well it can't be much worse than this. If it doesn't cool off, this whole marsh will dry up and all the fish will die," I said, pointing to the big expanse of green swampy land behind us.

"That's for sure. The mud's already starting to split and crack apart in places and there ain't but a little bit of water sitting in the bottom of those ditches," Leon replied sadly. "A lot of fish and stuff's going to die for sure, but this marsh just dries up like that every now and then, and there ain't nothing you can do about it."

We stopped at an oyster bar that was becoming exposed by the low tide and stooped down to pick off crown conchs, *Melongena corona*, and then went on to herd up some fiddler crabs down the shore. As we were wading around the soft spongy marsh grass that covered the land, we found some long-decayed pilings that were sticking up just above the mud. They were on the boundary line of our property, and I had wondered about them before.

"There used to be a big fishing dock here," Leon said, kicking out at the worm-eaten pilings. "George Henderson had a little fish camp right here and ran fishing parties out 'most every day. He had a pretty good business back then, he had at least six skiffs tied up. This dock went way out into the bay. It was almost as long as yours."

"How long ago?" I asked in amazement. I knew that wood was ephemeral and that sooner or later nature took over, but Leon was only in his early thirties. To see the very tips of posts sticking out above the soil, to tell the tale of another man's attempt at a business and his extension into the sea, was a bit of a shock.

"Let's see . . . I was about twelve, so that must have been twenty years ago. Most of this marsh grass has grown in since then. Used to be that this was a sandy beach and people would come from all over to swim on it, but this bay is silting in so bad that it's all changed. That big oyster bar that we just got those conchs off, that used to be an old barge that was made into a floating restaurant. It just kind of rotted away after it sunk and got eat up by worms and the metal rusted out and all them oysters grew on it. Now it ain't nothing but one big oyster bar."

"I thought that was natural!"

"Hell no, it wasn't. That was clean sandy beach there twenty years ago. When they built the county beach, they were going to dredge it up and get it out of there so folks could have a place to swim. But old man Edgar Barnette, the man who ran the dredge, got the job half finished and quit. Like everything else around here, it got all messed up with politics and nothing happened."

I knew what Leon was referring to. The Panacea beach had been a sore spot with everyone in town. There it was, more than twenty acres of half-filled-in marsh, sitting there grown over with weeds and flooded by the highest tide, neither true wetland nor true dryland. Fifteen years before, an old lady had given it to the county with the understanding that they would build a public beach in memory of her son who had been killed in the war.

The county erected a monument to the young man, a concrete and bronze plaque, and cut down all the trees. They hired a dredge contractor to fill it in, and for months and months he pumped sand into the marsh while doing other jobs. The county officials were not satisfied with his progress or his price and declared at a heated county commission meeting that he was robbing them and taking more time than necessary. The contractor got angry after he collected his first bill and towed his dredge away, and the beach had remained the same ever since.

Sticking out of the mud flat, with oyster burrs growing all over it, was the sleeve of the dredge pipe. Marsh grass was creeping out toward it, and the sea was reclaiming its own. Leon pointed out where the pipeline had been and showed me where the dredge had been at anchor. "Edgar Barnette pumped out one hell of a hole out there. It must have been fifteen feet deep in low tide and it's still pretty deep, but most of it's silted in by now." He looked at his watch. "It's nearly four o'clock and we got a bunch of shipments to get to the post office."

I didn't return with him. I wanted to look around my marsh a little more and reflect a little. Over the years I had been much bothered by environmental problems and by man's abuses of the environment, yet I knew that nature had a way of taking care of itself. The scars were healing—slowly, but healing nevertheless. The filled-in beach was now completely overgrown with bright green spike grass, *Distichlis*, which grew in the highest reaches of natural salt marshes. Where the sand had been pumped in and piled up to its highest elevation, too high for salt marshes to grow, weeds and shrubby bushes were taking over. Over the years the tides came in and the tides went out, carrying the silt that had been stirred up by the dredge. Once the county beach had had raw white sand pumped up on it, but now there were fiddler crabs living there, and periwinkle snails on the short stubby growths of cord grass, *Spartina alternaflora*.

It was a long way from a natural wetland. The natural flow into the marsh had been cut off and now instead of a sheet flow, it was nourished by a large deep drainage ditch that separated my property from the county's. My two acres of marshland wasn't anything to brag about either. Its natural water flow had been interrupted when the county filled in their land and it too received its water from the ditch during high tides. Therefore it wasn't the healthiest, best-flushed wetland, but it was the only wetland I had. Almost every type of typical salt-marsh vegetation of north Florida was represented there, not in dense stands but in scattered remnants. It was more typical of an upland marsh that one finds at the head of a bay. It was transected by several fire-break ditches that had been dug through it several years ago and the ditches generally stayed full of water and had shrimp and killifish in them.

However, there were times when the tides went very low and almost no water would get into the marsh. Over the years I had seen periods when all the water would evaporate from the ditches and thousands of killifish, silversides, menhaden minnows, grass shrimp, and other creatures would perish and dry in the sun.

Because the marsh sat between two filled-in parcels of land and was nourished only by the ditch instead of a natural creek, the water level seemed to have nothing to do with the rest of the bay. I was puzzled by it. Sometimes when the whole bay was dry and the mud flats were exposed for miles, the marsh would be filled with water, like a giant sponge. You would expect to see it drained completely but fish were swimming around happily. Then when the bay was swollen with water and waves were lapping on the barnacles on our pilings, the marshes would be drained completely. It was all very confusing.

I often worried about the health of my marsh. We gathered fiddler crabs from it, and killifish, and it was important to us. But every year at election time there were speeches about finishing the county beach, pumping it up higher, making it bigger and better. Yet in all likelihood the silt and run-off would damage my marsh even further. The silted turbid water blasting out of the dredge pipe would probably smother much of the fouling organisms that grew on our dock.

We built a little wooden walkway that stretched out over our living boggy marshland. Fishermen who used the dock would stare at it and shake their heads. "What did you build a dock up there for?" they would ask. "You can't get a boat up there unless there's a storm or something."

"We built it so we can look at what lives in the marsh without getting wet," I told them, "and show people why they're worth saving."

For a real show when I had visitors, I had only to take a piece of bread down to the marsh and throw it into the ditch; the water would soon be roiling with the bodies of little fish, flashing silver in the sunlight as they grabbed the bread, thrashed their bodies to and fro, and made off with a crumb. Before long the ditch would turn into a feeding frenzy, and this would stir up the blue crabs from the mud and they too would come charging out with their claws extended, snapping and pinching in all directions.

The best way to see a marsh is to get out and walk around. You have to accept a marsh on its own terms, by wearing tennis shoes or boots and long-sleeved shirts even in summer to keep from being bitten by the yellow flies. You can go forth, wading into the muddy boggy creeks, slogging through the shallow water, and be rewarded by seeing, touching, and feeling more life crawling and swimming around than your mind can conceive of. Walking on a salt marsh is like walking on top of some gigantic living organism, and after a while you can almost feel it respiring under your feet. You can smell the organic smells of life and death, of living growing things and dying plants, which all furnish nutrients and life to the estuaries. Marshes are not wastelands, as many believe—they are wonderlands.

A walkway represents a compromise between coming in all muddy, salty, and messy and still having access to a marsh. Often as I walked out over the wooden stringers that were supported on fence posts, only a few inches above the tips of the blades of grass, I was aware that I was still an intruder into this wetland. I would send a great blue heron soaring with its loud raucous squawk as it fluttered angrily away. Why such a magnificently beautiful bird wasn't endowed with a sweet voice, I don't know, but what comes out of its long neck is enough to make you jump out of your skin. More than once I have wandered out on our floating docks at night, thinking about catching squid or swimming sea cucumbers, and have been scared out of my wits by a loud "Squawk, squawk . . . squa-a-ak." It sounds almost like somebody has somebody else by the throat and is wringing his last dying scream from him.

Herons and egrets simply love docks. They know they are great places to catch fish, especially floating docks. All they do is perch themselves and wait, and as the golden anchovies whirl and jump at the surface, they bob their heads down and come up with their razor-sharp beaks filled with fish. Pelicans also like to sit on docks and watch the water, and when there are no people about, the docks belong to the birds. The very planking on the dock provides food for blue herons that prowl around, spearing little black grapsid crabs, *Sesarma,* that scurry under the boards. Sometimes a flock of noisy fish crows takes the dock over, and occasionally a yellow-crowned night heron perches on the pilings. They stay for a time and then leave.

The marsh walkway was especially popular with birds when the water was drying up in the high marsh, shrinking the water in the ditches and trapping fish into puddles. Then birds would come from all around to have a real feast. It is a natural process: birds always feast upon fish that are trapped in a drying pond or puddle. As the water began shrinking in the ditches within my marsh, the true story of just how cruel the world could be was being told.

It was clear that there was going to be a massive fish kill before long. I had seen it happen before, in that very same ditch. When it did, I was so moved that I tried to dig it out deeper. I couldn't just shovel the dirt out of the ditch and throw it on top of the marsh, that would destroy the grass. I had to push the wheelbarrow down the walkway and dig the mud out from the bottom of the ditch, then throw it into the wheelbarrow and push it down the walkway to dump it on our fill dirt.

The deeper water helped; probably there would be instances when it would provide enough water until the tide could return and save the lives of the trapped fish, but this wasn't one of those times. There was less than three inches in the deepest part of the ditch, and the hot hazy day gave no indication that help was on the way. If there had been a strong rain, water would have filled the ditch. The marsh killies grass shrimp, mullet, and crabs could tolerate low salinities but they couldn't tolerate dry, cracked, waterless mud.

Had this been a normal wetland with natural slopes and drainage, every single fish and swimming creature would have been out in the creek, following the falling tide. But as I stood on my walkway, looking down into the ditch where the fish were flurrying and gasping for life, I knew that their situation was hopeless. Over the past year hundreds of thousands of fish had entered this swampy muddy ground to feed upon the rich organic muds and browse among the strands of salt grasses, and they managed to get out of the marsh as the water seeped out into the drainage ditch. During periods of very high tides out in Dickerson Bay, the water flooded over the berm of woody vegetation and swamped the marsh, and fish moved freely back and forth from the bay to the marsh.

But these fish were in the wrong marsh at the wrong time. Down in the ditch, the muddy water was crammed with little frantic bodies. I leaned down and scooped up more than twenty in my hand, and all the little fish flurried and wriggled and tried to escape. Even though the oxygen in the water must have been practically nil, and its temperature was so high that I felt uncomfortable with my hand in it, every one of the little fish was alive.

It was amazing how much stress these killifish could take. They would remain alive in the marsh until the very last drop of water evaporated. Some might remain alive dug down in the soft black mud where there was a hint of dampness, waiting, hoping—that is, if a fish can hope—that the weather would change, the south wind would blow and the tide would come rising up into the marshes.

I stepped off the walkway and walked over the drying muds. There were other portions of the marsh where water had been trapped and it was already too late. The puddles were completely dry and the mud was cracked, and the dehydrated bodies of sail-fin mollies and rain and striped killies were everywhere. There were signs of large birds that had been feeding in the mud, herons probably, with their characteristic three toes. They had eaten all they could, and there were signs that raccoons too had devoured their fill, and still there were dead fish in excess. Flies buzzed around their parched withered bodies. I scooped away the dampest portion of the mud and found bodies several inches down, and hidden deeply as possible were two small live blue crabs. The crabs would probably survive the drought. The shrimp had perished along with the fish. Everywhere were the opaque orange bodies of the little crustaceans which looked as if they had been dropped into boiling water and cooked for dinner.

When you look at a disastrous scene like that, you know that in nature life is cheap. It is produced in abundance and can be destroyed in abundance. I walked back to the shrinking pool where the backs of the fish were sticking out, and I looked at the relentless sun overhead. Should I play God and intervene? Was it really my place to interfere with nature?

Probably. This really wasn't a natural disaster that occurs after a hurricane or a flood, or a fresh-water pond that dries up, killing tadpoles, fish, and insects. If this marsh wasn't trapped between two filled-in parcels of land this wouldn't be happening at all. In all my years of stomping around wetlands, I had never seen a natural, unaltered marsh go through the torture of this one. Once I had witnessed hundreds of shrimp dying behind a seawall in Mississippi, but there the natural flow of the marsh had also been cut off and the broiling summer heat was taking its toll.

I looked at my marsh. Most of the time the ditch through the marsh was a refuge for tiny fish that moved into the grasses to feed. The ditch was man-made, dug years ago with a plow through salt flats and highland marsh. If it had not been for the hand of man, fish wouldn't have been there. I'm sure that in the past creatures benefited from it. But now they were trapped in this unnatural cut in the grass. If I hadn't dug it out further and made it almost a foot deeper, all the fish that were trapped would have been dead along with the others.

So in a sense I had already interfered, and using that reasoning, I was about to interfere further. I was going to get my plastic garbage can and my dip net, and start swooping up the liquid mud down there and all the fish and dump it into the sea.

At that very moment a yellow striped Gulf salt marsh snake slithered out of the grass and rested on the shrinking mud bank of the ditch. I froze. I had seen *Natrix fasciata clarki* before, but only as a ghostly form whipping through the marsh or swimming madly in the creek, and now there was one out on the mud flat, totally unaware of my presence. That is a great thing about a walkway or a dock, it allows you to get up above wildlife and observe it while the creatures are oblivious to your presence.

*Natrix fasciata clarki* live exclusively in the salt marshes of Florida, tolerant of brackish and even high-salinity waters. People who fish at the edge of marshy ditches often encounter them because they normally live in the salt berms among the high-tide bushes, making forays into the marsh. Because of their long brown bodies some people mistake them for water moccasins, but they are not poisonous. Neither are they good-tempered. Once when I ran one down and finally caught it swimming across a creek it managed to inflict several savage and bloody bites that were excruciatingly painful. The snake exuded a musky smell that reeked of rotting fish, and defecated all over me. It was best left alone in tile marsh and observed from a distance.

Watching this snake from my marsh walkway was fascinating. It had come to this scene of death and despair to forage, and it lay on the bank of the ditch among the shadows of the green *Spartina* grass, waiting. The snake was totally unaware of my presence and I was able to squat down a few feet above it. It lay there with its serpentine head slightly raised, its black forked tongue whipping out, sensing and testing the desperation and confusion of all those hundreds of fish stranded in the shrinking puddles, gasping for oxygen and life.

Suddenly the snake shot out into the muddy water and began striking madly to and fro. The little fish began leaping wildly into the air; the brown muddy water was a flurry and the water sounded as if it were being pelted by a rainstorm. Back and forth the snake twisted and lashed its yellow striped body, and then in a flash that was almost too quick for the eye to see, it struck a large sail-fin molly and undulated its way back to the bank to swallow it. As it lashed the water with its sinuous curves, cutting the agitated surface, it held its head up high out of the water, with the wriggling fish locked firmly in its jaws. I could actually see its teeth embedded in its hapless prey. I watched with fascination, almost hypnotized by the serpent, for there was something awesome and yet wonderful about it. When it slithered up on the bank it went into a loose coil under the grasses and slowly swallowed its prey.

The rest of the fish in the ditch settled down. They had survived the immediate threat, and now they were trying to survive their futile race against time, until that moment when the cool waters of the sea would come seeping back into the marsh, filling the ditch, overflowing its banks and inundating the grasses. But there was not a breath of cool air, no hint of a gust of wind to push the water up. Not a bush moved, not a leaf flickered, and the heat bore down. Even as I sat there watching the snake swallowing the molly, I was sweating and uncomfortable. Then another marsh snake joined the first at the waterhole. It was a smaller, brighter-colored fellow and it didn't hesitate. It slid into the water and started driving the fish into a panic. Adroitly it snapped its jaws down into the water and its fangs sank into a fat sheepshead killifish. The young snake crawled out on the opposite bank to finish its meal.

The larger snake was ready to feed again. I watched it curiously as it writhed around doing its dance; it seemed to prefer to use its weapon of terror more than the juvenile did. It almost luxuriated amidst the panic and confusion. I stood up and my shadow passed over the snake, but it didn't notice. I even waved my hand above it, but it was absorbed in this macabre scene. But for all the terror, all it did was grab a single fish and crawl back up on the bank to eat it.

Anne drove up to the dock and I called to her. She sat beside me on the walkway, watching the snakes intently. "I wonder how they normally feed," she mused. "This is certainly a bonanza for them, but they don't often find fish trapped like this."

"I don't know," I replied, "maybe they lie in wait and strike as the killifish come by. Maybe they thrash about and chase little fish up to the head of the creeks and frighten them out onto mud flats and then attack."

"There was a telephone call for you. Someone wanting electric rays or something. You're supposed to call back right away."

"I will, as soon as I turn these fish loose."

We stood on the walkway and, with a dip net that Anne had brought, scooped up the fish. The snakes fled off into the marsh as soon as we started. As we dipped, the water literally boiled with frantic fish that were even more frantic than when they were under attack by the marsh snakes. The dip net was heavy, because I was catching a lot more mud than fish. It was loaded with soupy liquid mud that was being beaten up by the continuous turmoil of the fish. There was also a heavy suspension of detritus in the bottom of the ditch, tiny pieces of fiber and fragments from decaying marsh leaves that had been their food. There was also a number of pinching blue crabs.

We dipped and used two hands to lift the net, dumped the living slurry into the garbage can, and dipped some more. I could see that at best I would be able to save only a few of the fish; it would have

been impossible to get them all. Perhaps it was only a token, but we still saved thousands.

The plastic garbage can felt like it weighed a ton. We lugged it down the wooden walkway out over the salt berm that kept the waters in the bay out of the marsh, and dumped it into the sea. As the thick slurry of mud, fish, and water poured out over the clean shallows, a large turbid plume spread out, and thousands of tiny frantic little fish, crabs, and shrimp spread out and swam into the dangerous freedom of the sea.

# 6
# The Approach of Doom

The next day I returned to the ditch. It had dried up completely and the bodies of the unsaved remained behind, drying in the sun. Clouds of flies buzzed around the dehydrated little fish, and in the days to follow even the blue crabs that dug down into the mud perished. Then, overnight, the bay was swollen with water, almost too much water, and once again the marsh was flooded.

With the returning tide, life came back into the marsh. Swarms of killifish and young blue crabs moved up into the shallows to feast upon the kill. What the water birds had failed to do, the hungry little mouths of fish accomplished. In a few days' time every corpse had been picked clean. Not a dead fish or shrimp or crab could be seen anywhere. It was as if the disaster had never happened.

The water continued rising It covered the salt flats that were growing over with pickleweed, it flowed in over the high berm where the *Iva* and *Baccharis* bushes were flowering with their fall white plumes. The water seeped through the highest needlerush marshes and almost wetted the roots of the pine trees and on the dock it completely covered the barnacles. The floating boat stalls were standing higher than the stationary dock, and we had to step up to get onto them.

"I wonder why the tide's so high?" I asked Leon as I walked along the floating dock, dragging my dip net through the water.

"Ain't you heard? There's a storm kicking up in the Gulf of Mexico. It's supposed to be off Pensacola someplace and might be heading our way. That's what's pushing all this water into the bay."

Except for the high tide, you couldn't tell there was any weather disturbance. The sun was still shining, the seas were calm, there was hardly a breeze in the trees. The white-flowering high tide bushes swayed slightly in the breeze, and everything looked all right, but somehow it didn't feel right. There was a stickiness in the air. For a moment I thought I was imagining it.

Then Leon said, "We're gonna have to watch this weather. I don't like the feel of it. This time of year a storm could sure as hell mess us up."

A large tropical depression coming ashore near Panacea, with howling winds, blasting rains, and flooding tides, could play havoc.

As I stood on my dock and watched the tide steadily rising in the bay, covering the salt flats and submerging the succulent salt worts and grasses beneath a sheet of water, I began to worry. The boats would have to be anchored or hauled out of the water if the storm really hit us. Then there was always the worry that my dock could be damaged.

We were already well into our business season in October. Day and night blended into one as orders came pouring in from the new catalog of marine specimens that we had circulated to schools over the summer. The telephone rang incessantly, this one wanting sea urchins, that one fiddler crabs, electric rays, or three pounds of frozen jellyfish. The mailbox was jammed with letters, telegrams, purchase orders, and I found myself spending eight to ten hours a day with Leon, Doug, and Edward hurrying from tank to tank, digging burrowing gray sea anemones out of the sand and putting them into plastic bags, looking for that elusive purse crab that we thought we had put in tank 13 or measuring the carapaces of horse-shoe crabs. When we weren't filling orders, we were out in the boat, dragging nets and looking at our watches anxiously because we had to get in and ship the specimens off to meet specific flight deadlines.

During October it was a harried business. The specimens had to arrive at their destinations on specific dates and often specific times, so that professors could use them in their classroom exercises. The schedules of whole laboratories were based entirely upon the organisms that we were going to supply. It was an awesome responsibility, and if we didn't meet it there was hell to pay.

When all the orders were packed off and sent to the post office or the airport, and the lab was cleaned up for the day, and I was exhausted, the dock served its most delightful function. I could collapse on the floating dock, lie there, and watch all the life around me. I could relax and talk to the fishermen who came down to cast their lines off the dock. The dock gave both financial remuneration and profit to the soul, and that's an unbeatable combination.

It had been more than three years since I had purchased the dock, and it had withstood rough weather before. It stood there day after day, baking in the sun, standing the freezing winter winds, its timbers and decking parching and turning silver. It sopped up rains and withstood the rough seas that surged in with the south and east winds. It could take all that, but it had never really been tested in a hurricane. I seldom worried about the dock, but I didn't take it for granted. When we were most busy, we depended upon it a great deal. The dock provided us with a financial blessing: no matter how much demand we placed on it by ripping up its fouling organisms or pulling blue crabs from traps that were tied to its railings, or dipping up jellyfish that moved by, it usually produced everything we needed and had plenty more to spare.

We were working on the University of Nebraska's order for fifty comb jellyfish, *Mnemiopsis macrydi*. It was a good order, easy to get, because they were glittering and flashing everywhere around the dock. We could have collected five hundred if we had wanted them. The high water was bringing them in in larger quantities than I had seen in months.

Yet as I watched them moving in the swollen current, I knew that in a few months, in the very dead of winter, they would probably disappear from the sheltered bays and move off into deeper, high-salinity waters and scatter to the wind. Then I would have to tell my customers that they were simply available. We couldn't collect a thousand and keep them in our tanks, because they wouldn't survive more than a week or two. They were to be captured, studied, and enjoyed for a short time only, and then they would fade like freshly picked flowers.

At night, ctenophore jellyfish are magnificently bioluminescent, giving off a fiery blue light from eight rows of beating comb plates. They are as clear as drinking water, except for the delicate little zippers that rhythmically beat, giving the little ball some power of mobility. In the daytime, the comb plates refracted the sunlight into the shimmering colors of the rainbow. Almost every institution wanted ctenophores to show to their students, yet they were merely a dab of protein and salts; they were ninety-nine-percent water. Holding one of these transparent jellylike globs in the palm of my hand, I wondered how they could withstand the raging seas, the sweeping currents, and the moods and hostilities of the sea. If we didn't handle them ever so carefully they would tear and tatter; they had to be gently removed from the sea and transferred to our buckets. Even then we lost many, and we had to spend a lot of time picking out the perfect uninjured specimens from the ones that had been damaged in capture.

We had learned a great deal about ctenophores over the years. We didn't learn in a controlled scientific manner, by focusing in on them and watching them day after day, measuring their reaction to the waves and wind. We learned pragmatically, motivated by the needs of our customers and our desire to fill their orders. When the University of Illinois wanted a dozen comb jellies, we hurried down to the dock to find them. Sometimes we did, and sometimes we didn't. When the sea was calm and there was hardly a ripple, we often found them floating lazily on the surface. But when the winds began to blow and the water to roughen, they sank down into the depths and dispersed. Often if there was calm water between the hull of *Penaeus* and the pilings, or between the floating boat stalls, the jellyfish would aggregate.

Ctenophores belong to a different phylum from ordinary umbrella-shaped jellyfish. Unlike the coelenterates, they don't sting, and they have a more highly evolved nervous system. To me there were few creatures in the sea as beautiful, yet I knew that there were few as deadly to all the tiny plankters that drifted about the estuaries. They could eat their bulky watery weight in tiny newly hatched fish in an hour, or could come spinning, twirling and flashing through a soup of larval oysters and consume them by the millions. But what were a few million larvae here or there? These creatures were so small that I would never be able to see them. Plankton is a whole universe unto itself. A drop of water from around the dock often contained glittering green phytoplankton cells, copepods, fish ova, arrowworms, barnacle larvae, and larvae of other creatures that I couldn't begin to identify. That wasn't just water that was welling up into Dickerson Bay, it was a living soup. Sometimes the plankton could explode in such numbers that it made the waters murky.

It was this living soup flowing past the pilings and floating docks that provided nourishment for the fouling organisms that attached to our dock. The little feathery appendages of the bryozoans expanded from their calcareous trapdoors to partake of the food that drifted by. The barnacles flung out their legs and grabbed at the tiny copepods, and the sponges on the dock pumped in water through their canal systems and removed food and delivered it to the worms and snapping shrimp that lived within.

It takes plankton-rich, turbid estuarine waters to support a heavy encrustation of fouling organisms. You don't find heavily fouled pilings in the crystal-clear waters. Only where the plankton is abundant will you find masses of sea squirts growing on wharf pilings along with the bearded growths of hydroids.

All these creatures begin as planktonic organisms that drift about the sea and settle down on hard surfaces. In the case of hy-

droids, growth is exceedingly rapid. Quite suddenly little buds begin to develop in the parent colony and spring from the middle of
the polyps. Some species branch out from the stems and turn into
jellyfish that break off, carried away by the tides and currents. The
little medusae begin madly pulsating their umbrella-shaped bodies
and fill the water with their tiny contracting and expanding glassy
bell-shaped bodies, which glitter in the sunlight. When the moment
is right, these lilliputian jellyfish release a shower of sperm and egg
into the water and the resulting embryos become creeping larvae
that attach to pilings, adhere to sponges, or are repelled by barnacles.

If a larva is successful, it grows into a large branching colony just
like its parent. Our dock was usually covered with pink tufted hydroids. *Eudendrium carneum* grew densely on the floating docks
during the summer and fall, and throughout the winter we found
*Clytia*, which bore elongated stems and only a few branches. Occasionally the stinging hydroid, *Pennaria tiarella*, sprang up from the
Styrofoam with its fernlike polyps. I had mixed feelings about this
species. It was by far one of the most beautiful colonies, with its dark
black stems and its pinkish white hydranths, but it was also extremely unpleasant to handle.

Years ago, when I started Gulf Specimen Company, I had some
large orders from biological-supply houses for *Pennaria*. Not
having a dock of my own, I dived around other people's docks to
find them. There was one dock in particular that had massive
colonies of this fern hydroid growing all over it. This was a very old
dock—it dated back to World War II and was severely eaten away
by boring worms and isopods and heavily encrusted with every
conceivable sea creature. When I put on my diving mask and swam
down beneath the surface, I was in a world of blossoming magnificence. The pilings were covered with color and form and shape.
There were encrusting sea porks, *Amaroucium stellatum*, jadelike
green fleshy clumps, colonies of the black *Botryllus* tunicates, which
had golden star-shaped zooids embedded in their tests. Growing at
the base of the piling, *Polyandrocarpa maxima* screamed out its

redness to me. There were yellow sponges, and white barnacles, and everywhere there were royal plumed branches of the much-desired *Pennaria tiarella*, with their pink reproductive buds that sprout out from the side of the polyps, shaped like tiny watermelons. The polyps have long knobby tentacles that are studded with nematocysts, but you need a magnifying glass to see all that.

The magnificence of that underwater scene increased when I saw a school of black-and-white-striped spadefish browsing among fronds, nibbling away at the small crabs and skeleton shrimp that lived among the colonies. There was so much color everywhere, greens, pinks, whites, blacks, bright reds, yellows—it was all so wonderful. Yet, as I have driven through busy cities and looked at the car-scapes I have often wondered why I find the same colors so offensive in man-made structures. The big neon orange hamburger sign, the yellow carwash sign, the garish purple trim on a fried-chicken store, all make me cringe and wish I were out in a soft subtle countryside.

Yet when I saw similar colors down below, I was enchanted by them. Perhaps the colors were softer down there. They didn't have flat, hard, shiny plastic surfaces that grated on my eyeballs. But the vivid purple tentacles of the sea anemones at the base of the pilings had the same purpose as the vulgar garish signs and eye-catching glitter in the cities: to attract passers-by.

When I was attracted to the beckoning plumes of the hydroids, clouds of nematocysts were discharged into the water and my body was covered with fiery welts. By the time I had completed my order, I was one mass of splotches and became quite ill.

Getting hydroids off our floating dock was a much more pleasant method of harvesting them. After Leon and I had dipped up all the jellyfish we needed, we attended to the zebra shrimp that lived among the hydroid colonies. We dragged our dip nets through the clusters as we walked down the entire lengths of the boat stalls, causing the black and white shrimp with orange-spotted legs to jump out and be swept into the webbing bags. Then we reached down under the dock and tore off clumps of green sponge,

*Halichondria bowerbanki,* that had tiny green algae cells living within the colony, along with hundreds of small pink sea anemones, *Aiptasiomorpha texaensis,* that were scattered over its surface.

We pulled off clusters of the pink hydroids, washed the mud off the branches, and filled more buckets to bring back to the lab. If we let them sit in the water and stagnate for a few hours, sea slugs would come popping out and flatworms would leave the colonies and climb up the sides of the buckets. It was little wonder that so many schools teaching invertebrate zoology ordered assorted fouling organisms from us. Nearly every phylum in the animal kingdom was represented in the life we scraped off our floating dock.

When a teacher unpacked our Styrofoam boxes and pulled out a plastic bag filled with hydroids and spread it out into fingerbowls of sea water, his students could actually see the unbelievable density and diversity of marine life that could grow so rapidly. Small brown snails, scarcely a quarter inch in length, would protrude from their shells and cling to the stems and branches. Almost microscopic larval clams could be found among the polyps, and there would be clouds of copepods and other minute crustaceans orbiting in the colony like moths flying about a light. There might be flattened porcelain crabs and tiny mud crabs and sea spiders that blended in perfectly and relied on camouflage to protect them from browsing fish.

The teacher could cut the colony into pieces and give one to each student to put under his own dissecting scope. And when the hydroid colony came into focus, the student might witness the tiny orange nudibranchs, *Eolis,* sliding and slipping delicately over the bushes of pink polyps, waving their purple projections and frills like a Spanish dancer. So perfectly do they blend in with the pink polyps that they might be missed unless the student looks very carefully. The sea slugs lay ribbons of tiny white eggs and cement them to the branches and then go about foraging among the polyps.

Nudibranchs are snails without shells. Some of them are large, bizarre, colorful creatures that catch the eye immediately; others, like the eolids, are small and live exclusively in the dense jungle of stinging flowers, deriving both protection and food at the same time. Like caterpillars devouring the leaves on a maple tree, the nudibranch munches away on the contracting and expanding hydroid polyps, methodically chewing up and devouring the piercingly stinging tentacles. No one knows why or even how this little sea slug ingests the painfully stinging nematocysts into its own body and magnifies the power of its prey, but it does. Certain primitive peoples in remote jungles believe that if they eat the heart of a lion or some other strong and courageous animal, that courage and strength will be absorbed into their own bodies. In the case of this tiny naked snail, that really happens.

When the student first looks at the colony, the whole thing may appear to be writhing and alive. What he is seeing are thousands upon thousands of tiny brown skeleton shrimp, *Caprella acutifrons*, which blend and flex their bodies, forever moving up and down the branches. Watching them closely, it's amazing what peculiar creatures they really are. Hollywood couldn't dream of a better monster. If they were only two feet tall instead of a half inch, no public aquarium would be without one. Their antics would put porpoises to shame. Almost as if they are praying, they bow slowly down, clasping their palmlike claws together, and sway from side to side.

The minute skeleton shrimp are attached to the colony by clinging hooks at the ends of their bodies and they are forever foraging among the branches and polyps, scraping off bits of debris and removing diatoms. Now and then they'll start actively climbing around the branches, swinging from limb to limb like monkeys. First they bend down into a loop, take hold of the adjacent branch with their front claws, and release the hold of their hind legs. Then they swing their entire bodies over and their terminal hooks grab for a new hold.

In this manner these peculiar-looking amphipods constantly move about the colony, cleaning and picking. Added to their acrobatics is a strong maternal instinct. The female carries the eggs and larvae in a brood pouch that makes her look decidedly pregnant most of the time.

Fish love to eat caprellid amphipods. Often we would tear off a clump of hydroids, toss it in the aquarium, and see even the most finicky reluctant feeders go wild and gobble up the tiny crustaceans as fast as they could pick them out of the hydroids. Sea horses especially love to eat them.

Many fish are dependent upon the hydroid community for food. Schools of silvery flashing menhaden and silverside minnows dart among the tufted growths, feeding upon diatoms, bacterial and organic muds that are trapped in the colony. Then schools of black-and-white-striped spadefish move in to nibble away at the skeleton shrimp, tiny crabs, polychaete worms, and larval clams. Of all the fish that visit thc dock, the spadefish are perhaps the most beautiful and graceful with their broad black and silver stripes and their flattened bodies. Many people incorrectly call them angelfish because they look like fresh-water tropicals in the aquarium.

Many of the large public aquariums around the United States had standing orders with us for as many as one hundred spadefish in a single shipment. They made magnificent aquarium displays, because if one moved they all moved, in perfect formation. If you disturbed them, they would scatter off in all directions, but it only took a moment before they regrouped.

Only in the early fall did they appear around our pilings in significant numbers, and when they did, we baited our traps and used every available method to catch them, including throwing cast nets.

The dock was a good revenue producer, there was little argument about that. We grew creatures off it, netted animals that drifted by in the currents, trapped and even harvested food from it.

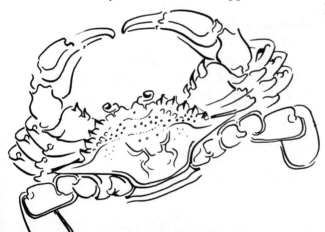 Large edible fish such as flounder, sheepshead, grunts, drums, trout, redfish, and sometimes mackerel could be caught on hook and line by just casting out into the channel. During the

winter, large fat succulent oysters grew on the pilings and oyster strings that we hung off the dock, and we could sit there in the warm afternoon sun and pry the valves apart, sprinkle it with hot sauce and swallow down the tasty meat, and there was nothing better.

We could always catch blue crabs, too. We had sales for a dozen here and there, but there were far more crabs available for mundane and gourmet purposes than teaching and research. When the traps were baited and the weather was warm, big handsome male crabs left the marshy creeks in Dickerson Bay and foraged around the bottom. They found their way to the traps sitting on the bottom baited with fish heads, with all those good smells emanating from them.

Then the crabs tried to get at it. Frustration upon frustration because the smell was so good, but the wire would keep them out. The crabs would crawl up and down the cages, climb on top of them, and then suddenly they would find a door and squeeze through it, just as if they were crawling into a small cave or burrow. There was the food sitting in the bait cages; they latched on to it and shredded it with their claws. But when all the food was eaten and the bones were picked clean there was no escape. The funnels that led them into the traps did not lead them out.

In the morning or two or three days later, when we lifted the traps they would be full of blue crabs. Often there were stone crabs with their massive grinding claws that were so prized by gourmets. But in addition to what edibles we gathered from the traps, there were always crown conchs, toadfish, and other fish to add to our inventories of living specimens.

Leon and I loaded our buckets into the dock cart and began pushing it toward the shore. We stopped, hauled up a trap, and emptied four spadefish out of it. "That's twelve dollars, right there," he said proudly. "This dock doesn't owe us a dime! It's the best investment Gulf Specimen Company ever made. Why, there isn't a month that rolls around that it doesn't make its monthly payment to the bank and then some."

It was certainly true. I once kept track of everything shipped in a single day, and found that the dock had contributed almost a hundred dollars' worth of specimens to our orders. It was collected without the chagrin of boats, gasoline, dredges, and all the worries and hardships of the open sea. But when Leon began saying how good the dock was, or how great *Penaeus* was running (although he had been saying it less and less), I became nervous. I suppose I am a superstitious person. I come from a stock of medieval Europeans who insert thumb and forefinger together to make a "phaig" to ward off the evil eye. There are jealous gods or demons lurking out there, and somehow deep in my cultural background I have come to believe that if you brag on something or extol its virtues too much, you are inviting the glare of disaster.

Listening to Leon's enthusiasm, his bragging about how it was the best dock in Panacea and how no one had such a fine structure anywhere along this part of the coast, I couldn't say, "Hey, Leon, hush up. What are you trying to do? Invite the glance of the evil eye?" That would sound silly.

"Now you show me a business that has the *profit* this one does. Sure, I know we ain't making no money 'cause we got big expenses and all. But who else can put a few oyster shells on a string, hang them off the dock and pull up two hundred dollars worth of stuff? If we just had more sales for this stuff, we'd be rich!"

Again, I had that sinking feeling. It usually went away, but this time it didn't. The air was sticky, not a leaf moved, but there was something wrong, something ominous. Leon quit bragging about the dock as we pulled the cart to the truck and began to load. He looked out over the bay. The sky was suddenly getting darker. "I'll tell you, this is funny weather. I sure don't like the look of it," he said.

That evening I flicked on the television and learned that the tropical depression was moving our way. It was by no means a true hurricane, the weatherman said, it was one of those big, undefined circular wind patterns. By the hour the weather began to deteriorate. It got darker and darker, the winds began to howl, the rain started pouring down. Wind-driven rain whipped down on the little village of Panacea, and darkness crept in with it. I went to bed with the sound of thunder rumbling, lightning splintering, and rain beating down upon my roof.

I hoped that by the next morning it would have blown over and there would be peace.

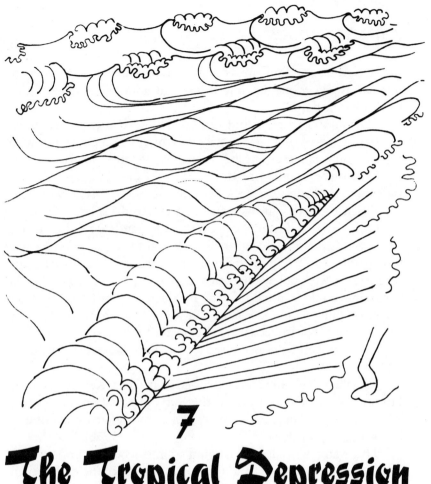

# 7
# The Tropical Depression

But the next morning brought no peace. During the night the storm had intensified and the winds increased. Instead of warm friendly sunshine beaming through my windows, there was only a dull light and the monotonous downpour of rain beating upon the tin roof. I turned on the television and listened moodily to the brightly smiling weather girl talking about rain, rain, and more rain. A real cold front had developed. A big mass of chilling high-pressure air was moving down from the north and would meet the warm tropical depression out in the Gulf. "You know what that means,"

I said to Anne as I drank my coffee. "We're going to have a mess on our hands. We won't be able to get out and collect anything for three or four days."

"I wouldn't worry about not collecting. If the seas get much worse out there, we may not have anything to go collecting with. I hope you put something between *Penaeus* and the dock."

"We put a couple of extra rubber tires on her side. She was starting to scrub up against the dock last night but I'll check on her."

When I looked out over the bay I received a shock. The dawn light revealed that during the night the water and tide had risen. The south winds that had been howling all night had pushed the ocean waters into the bay and the marshes were nearly covered. Only where there were tall needlerush marshes and some fringes of green cord grass could you tell that the marshes even existed. From our house on pilings I could see the salt water inching up on the roots of the slash pines and cedars at the head of the marsh. All the high salt flats that were seldom wetted by the high tides were covered by several feet of water.

"Small-craft warnings are extended from Cedar Key to Pensacola, Florida," said the television weather girl. "Tornado warnings have been posted for Gulf, Franklin, Wakulla, and Jefferson counties as the storm approaches."

The water was already lapping at the wooden stringers and planking under my dock, and the upper reaches of the dark brownish black pilings and their tarnished silvery bolts, which had never been wet by the sea, were now under water. The waves broke under the planking, sending up a frothy white spray through the cracks. In the dawn I could see the outline of *Penaeus*, riding the waves, high above the dock, next to the railing.

The rain beat down and the wind gusted over the bay, and the sky brightened only slightly. The normally calm sheltered bay was frothing up into whitecaps and I began to worry even more. In six hours it would be high tide and then the water would really rise. The whole dock would be under water.

"Coastal residents in low-lying areas should stay tuned to their radios and heed evacuation orders," said the weather girl.

The rain began to pour abruptly down with tremendous energy, blending the air and the sea into one vast misty mass. My vision was obscured as thunder growled and lightning split the gray sky, revealing the darker clouds in the distance. I finished my coffee while watching the ominous clouds, wondering if a tornado really was going to develop. Suddenly it was all very real. I didn't fear the high water so much—my dock could take it—but if there were violent hundred-mile-an-hour winds bearing down with it, we would be blown away.

Suddenly I was not the twentieth-century superman who was the master of the environment around me, I was but one more frail creature living in a little wooden shack on pilings that overlooked the bay. All this time we had enjoyed its assets, its magnificent sunrises, seeing porpoises bobbing around catching fish, big plumy egrets stalking the mud flats, and we thought how lucky we were. Now there was the sea out there, the great gray angry sea that had swollen up in the bay, and I was worried. Where do you go when a tornado strikes? Can you crawl into your burrow, crawl under a rock, hide in the branches of a tree? There is might and power out there, tremendous energy that could destroy us and everything we had, and there was absolutely nothing we could do about it.

Yet there was satisfaction in that feeling. It was comforting to know that there was something out there that the U.S. Army Corps of Engineers couldn't master. There were forces beyond them. If the Corps wanted to, they could build a huge dike across the entire mouth of Dickerson Bay, pump out the water, and plant crops on the mud flats the way Holland has pushed back the sea. They could change the course of entire rivers, blast down huge mountains, but there was no aerial barrier that they could erect in the sky, at least not yet. For the present time, they couldn't corral the clouds and harness the winds and allocate rains the way they could open floodgates and control the flow of rivers.

I looked at the *Iva* and *Baccharis* bushes on the shoreline whipping back and forth in the wind. They beat with such fury that their white blooms looked as if they would shake off. The towering pines swayed back and forth, vibrating their needles as the wind cut through them and the rain showered down. At the end of our dock, the Styrofoam floats were bouncing and jostling as the waves increased and I wondered how and if they could survive the storm.

Even though they were alive with teeming multitudes of fouling organisms, they were still man-made. They were only wood and Styrofoam and steel fastened together with bolts and nails. No structure, no matter how rigid or strong, could withstand the bending, flexing, and rolling of the waves indefinitely. If they didn't tear apart when the tide got too high, they would rise over the pilings that held them in place and float off into the sea.

I looked at the marshes when the rain had died down. Once again they were a source of amazement. While the waters were high and raging out in the bay, far up in the marsh it was calm and slick with hardly a ripple. The grass absorbed all the wave shock, it took the surging energy away from the sea and made it quiet. But it was quite another matter where the waves lashed against the parking lot that had been filled in so long ago. There the waves fought against the hard-packed fill dirt, whipping it up into solution and carrying it away, leaving a washed-out, chopped-off cliff.

Rain water was deepening in the parking lot and when the puddles became swollen, they drained off into the bay, carrying the muddy, silted waters with it. The waters didn't run straight off my parking lot. They were eroding a gradual, roundabout curving flow that one finds in tidal creeks or rivers meandering to the sea, oxbows and all. Water is not meant to flow in straight lines the way a logical, mathematical engineer's mind would have it. It was meant to meander free.

I couldn't help getting an ironic bit of satisfaction from watching my valuable waterfront property erode slowly. Even though the needlerush and cord-grass marshes had been buried years before by dump trucks and dredges, long before I bought it, it would only be a matter of years before it would all wash away to the sea. The rising waters of the bay would one day reclaim my parking lot and turn it back into green marshland. The waters immediately adjacent to the parking lot and partly around the dock pilings had become silted and white from the limestone and dirt in the runoff and were staining the rest of the dark waters. But that silt would pass away before long; it would settle out in the bay and would build up the shoreline and mud flats someplace else, gradually allowing the green marshes to extend out even farther, making it more green and beautiful. The shoreline, the marsh, and the sea were part of a living, flowing, dynamic process.

I had seen so many beautiful areas in Florida fall to the bulldozer, the dragline, and the hydraulic dredge. I had talked myself hoarse at public hearings and had written hundreds of letters trying to save these areas, only to see meager compromises here and there, and the destruction commence. Perhaps it was a bit peevish on my part, but it was nice to know that nature had a way of reclaiming her own.

The floodwater rising out in the bay and the rain beating down on the land were part of a dynamic process of nourishing the estuaries. If man got in the way, man would have to take the consequences. As the water churned up into the highest reaches of the marshes, all the decaying grasses, purple with nitrogen-fixing bacteria and rich in protein, vitamins, carbohydrates and potassium, would be exported out to sea. Far from shore, perhaps in the depths of the ocean, the nutrients from this interface of vegetation that grew between land and sea would be nourishing plankton and giving food to filterfeeding organisms. I stood at my window, watching the enormous productivity of the oceans and marveling at it.

I heard our truck bouncing along the sand road, splashing through the puddles, and I saw its headlights.

"Get dressed and help me move these boats," Leon hollered unphilosophically up to me. "When the tide comes in in a few hours they'll tear the dock all to pieces."

I winced anticipating the unpleasant gusts of rain, cold wet rain, beating on my face and soaking through my raincoat. But the sea was calling; she was giving us her warning and if we did not heed it we'd be in trouble. Survival was at hand, so I had to forget the philosophical and get down to the practical. That wonderful dynamic process of storms and tides and sea, of nutrients, energy, food chains, and productivity, could ruin me.

There was light enough to see now. The rain was beating on my face, but I could see the boats jumping up and down from their moorings. The tide was coming higher.

*Penaeus* seemed to sense that she was in grave danger. There was none of her usual nonsense and crankiness. Leon no sooner turned the key than she cranked off with a mighty roar, blew carbon and smoke out of her stack, and sat there idling noisily.

When I tied the tunnel boat to her stern and cast off, Leon shifted the little shrimper into gear and churned along, plowing down the choppy waves, and soon we were several hundred feet from the dock. Leon turned the boat over to me and threw the anchor. I backed up until we felt it dig deep into the mud, then he hurried to the stern and dropped out the other anchor. He moved agilely about the craft, oblivious of his big yellow slicker suit, which seemed to swamp him.

When the engine died there was silence in the bay, only the sound of rain splattering on the sea and the rumble of thunder in the background. Suddenly the sky split open with a blinding flash of lightning and I shook in my boots. Leon looked up, startled. "Damn, that was close, Jack. Let's get the hell out of here."

We raced for the dock, never-minding the seas breaking in over the bow of the tunnel boat, which balked and cavitated in the frothy waves.

Now that it was as light as it was going to get, I looked around the bay and saw fishermen bailing out their boats. A number of them had pulled up anchor and were moving away from the city dock. Some were running up toward the head of the bay, where it was sheltered. They would take their boats far up a winding creek to ride out the storm.

As we worked to secure the tunnel boat, tying down its gill net

and putting away assorted loose buckets and gear that might be blown away, the tide was rising. Already water was slapping hard at the planking beneath our feet, splashing up through the cracks, and the wind was intensifying. All of Dickerson Bay was flooded now. The water covered almost the entire Fiddler's Point peninsula and still it was rising. Our little backwater bay had turned into a raging sea with five-foot waves of steel-gray water cresting in whitecaps and crashing down upon my parking lot, expending their wind-blasted energy in foaming, frothing turbulence.

These same raging seas would have looked quite normal along the outer beaches of the Gulf of Mexico or the Atlantic Ocean, where the open rolling sea met the shore, but here in this normally quiet shallow little bay they were quite fearsome. The high tide bushes with their little waxy green leaves bunched together on the tufted branches looked as if their long flexible stems were about to snap off from whipping back and forth so violently.

The rain started beating down even harder than before, and it was perfectly clear that this would be a day of filling orders, cleaning dead animals and shells out of the tanks, working on the air blowers, and maintaining the outboard motors. There was usually plenty to do indoors.

I looked over the orders that had been all neatly decked out, and groaned in dismay. There were a number of creatures due today, and there was absolutely no hope of getting them. One customer wanted two dozen brachiopods for a class project, and we didn't have any. Another wanted ctenophores, and even though the fragile comb jellies had filled the bay only three days ago, I knew there would be no hope of finding them in this weather; they dispersed in the winds and sank into the depths. But there were a few amphioxus left in stock, and we probably had enough sea urchins to get by.

Our order desk was next to a large round concrete tank where Little Bit and Susie, our two Atlantic ridley sea turtles, lived. Seeing me, they both swam over with their mouths open, getting as close to us as they could, begging for food. We often fed them by hand, but I didn't feel like bothering with them, so I just tossed a live blue crab into their tank and watched them take off after it. Ridleys love to eat crabs.

Anne was in the lab, filling an order from the University of Wisconsin for two dozen small horseshoe crabs. Horseshoe crabs, *Limulus polyphemus* aren't true crabs, they aren't crustaceans; they belong to a group of arthropods that are closely related to ticks, scorpions, and spiders. She was engrossed in watching them run over the bottom of the tank in their half-swimming, half-walking bobbing gait. Gathering momentum like an airplane taking off, they would lift off by beating their gill books and legs in unison, and arch their shells and swim up to the top of the tank. Then they would do a backflip in mid-water and begin swimming upside down on the surface, kicking their legs and beating their gills in unison like some sort of peculiar internal-combustion engine.

After a while the crabs would cease swimming and either flip over in the water or sink to the bottom and land on their backs. They would arch their jointed shells, thrust their pointed tails into the sand, twist sideways and flip over, and then proceed to run along the bottom or burrow out of sight.

"That really is strange," Anne said, poking one and making it sink to the bottom. "Why do you suppose they only swim upside down in captivity?"

She had asked herself that question many times before and still there was no answer. Anne had been working on her doctorate for

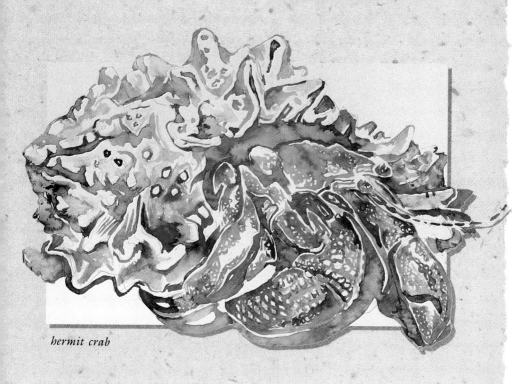

*hermit crab*

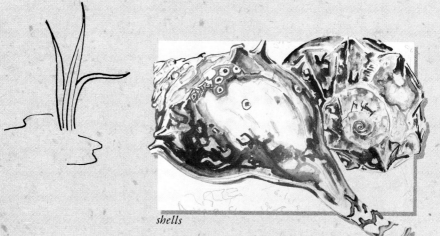

*shells*

*shells*

*speckled trout*

*frogs*

*turtles*

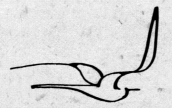

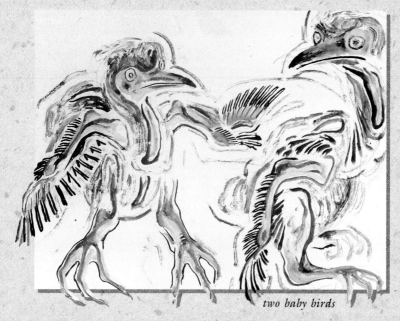

*two baby birds*

*hermit crab*

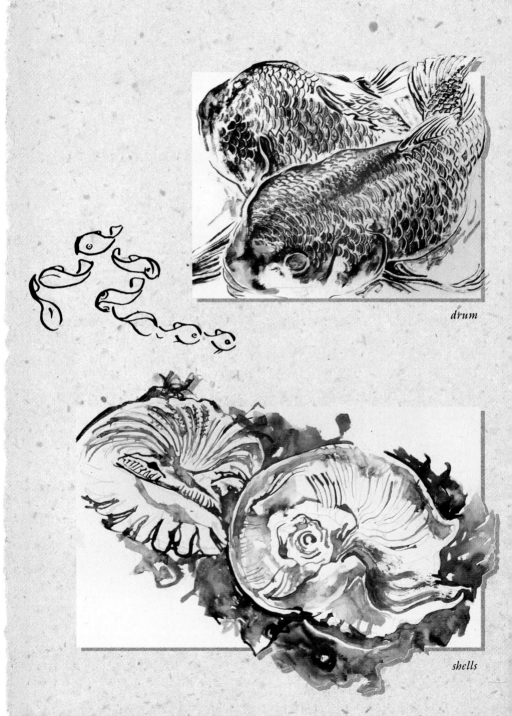

*drum*

*shells*

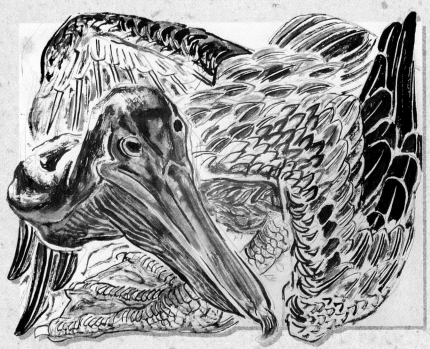

*brown pelican*

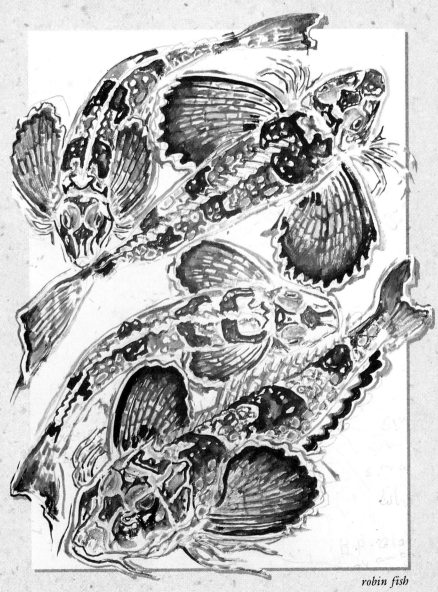

*robin fish*

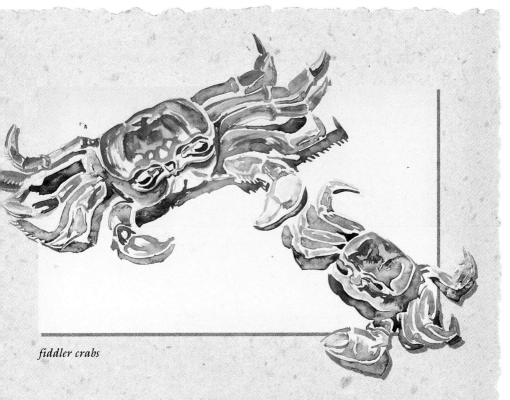

*fiddler crabs*

*turtle*

*flying fish*

*crab and starfish*

baby herons

*baby pelicans*

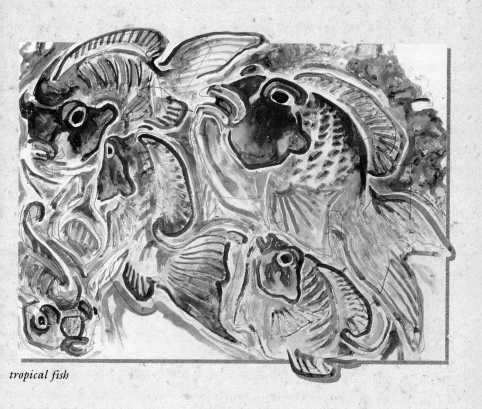

*tropical fish*

*hermit crab*

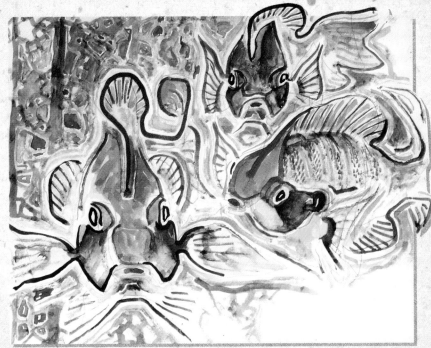

*three fish*

*purple shells*

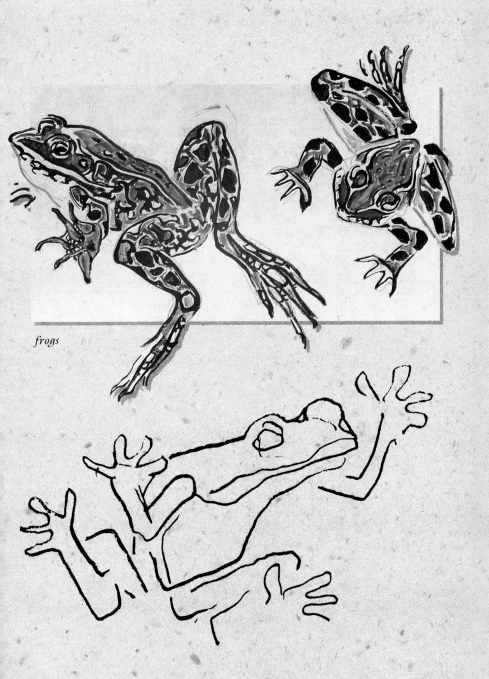

frogs

*robin fish*

*hermit crab*

almost a year now, studying the behavior of horseshoe crabs. She had spent hundreds of hours out on the tide flats observing juvenile *Limulus* make trails in the sand. She had watched the adults come up on the breeding beaches to lay their eggs and had observed them while diving, and not once had she, or I, or anyone we had ever talked to, witnessed even a single specimen swimming upside down in the ocean.

"Maybe they're protesting," I offered. "Maybe it's their way of saying they don't like being kept in captivity. Maybe it's an appeasement gesture, like a dog turning over on its back when it doesn't want to fight."

"That's ridiculous," she said, laughing. "Horseshoe crabs don't make appeasement gestures."

"All right, what's your solution then?"

"I haven't got one," she replied, picking out some small crabs and putting them into her bucket, "but no answer is better than that one. What are you after?"

"Two dozen red cleaning shrimp," I replied, and then grinned at her. "They swim upside down, too."

"Well, they don't do it to appease the fish. They're just displaying so that their host won't eat them," she retorted. At one time Anne had considered doing her dissertation on the cleaning behavior of the red cleaner, *Hyppolysmata wurdmanni*, but dropped it in favor of horseshoe crabs. We always tried to keep our aquariums stocked with red cleaners because they were so much in demand. They lived under the yellow boring sponges on tide flats and with their bright scarlet colors in longitudinal stripes and their distinct Y-shaped white markings on their undersides, they were beautiful little jewels. The coloration and their peculiar displays of rocking back and forth and swimming through the water upside down in a most unshrimplike fashion warned predatory fish that this was not a morsel but a friend.

We often put spiny boxfish into the aquarium and delighted in watching the red cleaners display and start their elaborate cleaning symbiosis. First they would begin their little dance of rocking back and forth, back and forth, prominently displaying their white thorax markings. Then when the fish showed the right kind of interest, which was to hover above them and not make any hungry lunges, the shrimp would leave their cover and ascend the vertical column of water. Sometimes they would take long stepping movements with their legs as if they were climbing a flight of stairs as they advanced down until they cautiously landed on the fish. Then, while the host remained perfectly motionless, they would begin picking and tearing at its parasites and fungus-ridden tissues.

I found that if I placed my hand in the aquarium and didn't make any sudden motions, they would ascend the water column and clean my fingers. What a peculiar sensation it was to have the little shrimp swarming over me, reaching their sharp little nippers down into my oyster and barnacle cuts, probing and picking and tearing at the infected tissue. Sometimes they reached down into the wound, got hold of some tender flesh, and I would let out a yelp. But invariably the wound would be picked clean and would heal perfectly without infection.

We had experimented with putting a variety of fish into the tank. The slow-moving spiny boxfish, the filefish, and the toadfish readily accepted the cleaning, but the voracious smooth puffers and rock bass looked at the shrimp only as food. They couldn't distinguish the cleaners from the grass shrimp that we fed them.

Leon came over and watched me swooping the red cleaners up in a fine-meshed aquarium net. "I'm damn sure glad we stocked up on them while we had a chance. I'll bet you that after this rain is over, we won't find any within ten miles of the shore. That fresh water will ruin collecting around here, you can bet on that."

But there were bigger problems facing us. While we were packing, the old derelict shrimp boat *Isabel*, owned by the Roberts Fish Company, was rising up from the dark rainy waters like some sort of ghostly monster from the depths. I had lived in Panacea for eight years and I had never seen the *Isabel* in good running condition, although fishermen used to talk about what a good boat she'd been before Roberts had bought her and let her run down.

On her faded cabin there were only remnants of gray paint, and her rigging was rusty and worn. It was plain to see that she was a vessel of the past. She had paid for herself and her owners didn't want to put in any money for repairs. If she used oil, they just poured her crankcase full of burnt motor oil, and if her cables broke they didn't get new ones, they just tied knots in the old ones and spliced them.

It had been years since *Isabel* had gone out to sea. She stayed broken down most of the time, and then one day after there had been intense rains, she sank at the dock. After a few weeks her owners raised her up, but she sank right down again and this time they left her down.

There she sat, a hazard to navigation, rusting away. Then after a few months George Roberts and his sons pumped her out, stripped off all her winches and everything that was worth anything, and pulled her up on a mud flat, using a car wrecker at high tide. They abandoned her, and there she lay for two years. The tide went in and the tide went out and *Isabel* was claimed by the sea. Her hull was bearded with sharp oysters and barnacles. Sea roaches scurried over the top of her hull, and gulls perched and covered her silvery gray weather-beaten boards with white droppings, which parched day after day in the sun.

But now with all the rains and rising water, the *Isabel* rose on her final journey. Like a ghost ship, all sixty feet of her lifted up from the mud, and she began to drift. Only her big open hull remained now; her rusty mast had been broken off and the roof of her cabin had rotted away. As the tide rose higher and higher and the winds came rushing down the bay, *Isabel* started drifting right toward my dock.

We were digging through the filter beds in our tanks trying to locate six blood clams when one of my neighbors who lived down the street telephoned that the *Isabel* was headed right for my dock.

I called everyone together—Leon, Doug, Edward, and Anne—and we hurried down to get a look at her, and there she was, that barnacle-covered floating wreck, looming ever closer.

"You'd better get old man George Roberts and tell him he'd better do something about his boat," Leon warned. "The rest of us will stay here and try to fend her off."

The dock was now completely under water. Only the high white-painted railings sticking up gave any indication that a dock existed beneath the gray waves. I could envision that encrusted derelict crashing into my pier, splintering the planking and bending the pilings. Weighted with water the way she was and powered by waves and winds, there would be no stopping her.

I hurried up to the Roberts Fish Company and found the Roberts boys unloading a truck of mullet that had just been shipped in. They nodded to me as I drove up, a suspicious nod. Over the years we had developed at best an arm's-length relationship, and Wilbur, the fatter one, finally drawled out, "How you doing, Jack?"

"Not very good at all. Your boat, the *Isabel*, is floating down the bay and she's heading for my dock. If it hits, it will tear it all to pieces."

"It ain't our boat no more," Wilbur said, shaking his head vigorously. "We sold to Sidney Wilson about a month ago. You ought to see him about it."

"He told me that you never transferred title to it and that he said he wouldn't have it. That makes it your boat, Wilbur," I said angrily, "and if it tears my dock down, someone's going to build me a new one."

"It ain't gonna be us. That's old Wilson's boat."

"I'm not going to argue with you about it, it's coming down on my dock right now," I said, gritting my teeth and turning toward my car.

George Roberts came out of the office. He had apparently overheard our conversation. "Like my boy says, it ain't our boat. But we'll be glad to be neighborly and help out. Is there anything we can do?"

"Yes. Go get your boat and tow it away from my dock, or anchor it, blow it up, do something—anything. Just keep it from tearing down my dock!"

He ran his finger over his unshaven face. "Our big boat's over in Carrabelle, and that little skiff boat went to the bottom last night,

Wilbur tells me. I ain't been down to look at it yet."

"You mean that's your skiff out there? The one we saw this morning with just the motor sticking up?" I asked incredulously.

"Yeah, the boy let it rain down. You know how it is. . . . We let that boy Roy Suthers use it to fish mullet for us, and he's the sorriest thing . . . he don't care about the other man's equipment. You know yourself, you work these boys around here and—"

"George, I've got to go. Come on down and help us push it off, maybe we can tow it with our boat, or something." When I returned to the dock, the water was at peak tide and all the marsh grass was completely flooded over. It was one vast gray body of water. Only the very tops of the high tide bushes along the shore stood whipping in the waves, and the road leading down to our dock was inundated. The bay was full of debris, old fish boxes and garbage that had been discarded and long ago washed up into the marsh, where it had sun-dried, and now it was suddenly floating again.

And there, most impressive of all amidst the debris, came the Isabel bearing down on the tip of my dock.

"She's coming mighty close," Leon called to Wilbur as they waded over the planking, "but if she keeps on she may miss it by two or three feet."

"I sure hope so," wailed Wilbur, who could really see the magnitude of the hulk, "'cause we don't want no trouble."

"We ain't got no trouble," snapped George. "Like I told Jack, we done sold the Isabel to Sidney. You know that yourself, Leon."

"No hell I don't," retorted Leon, watching the half-sunken hull coming ever closer. "I talked to Sidney in the café yesterday, and he said he wasn't messing with it. He ain't paid you no money for her and ain't signed no papers."

"But he said he bought it," Wilbur replied hotly. "A man's as good as his word."

"Not when it's going to cost him two or three thousand dollars or more to fix a torn-up dock. Or it will if we don't shove it or something," Anne said, as she leaned on the rail.

"You can't shove that mess off," Leon retorted. "She must

weigh twenty tons filled with water. You'd cut yourself all to pieces on barnacles if you even tried."

The wind started to gust harder, and the waves were breaking over our knees. "Edward, Doug, go down there and bail out the tunnel boat and get it ready. We'll get *Penaeus* and tow that old piece of junk out of the way. Maybe the tide will turn and she'll tear down the city dock."

"No, that tide's still coming in," said George. "It'll carry her right up to the head of the bay, where she'll go aground on a mud flat and be out of everyone's way."

"Maybe," said Anne, "but she might come right back on top of us and tear our dock up in the next storm if she doesn't do it right now."

While Edward and Doug were bailing out the tunnel boat, pushing it off the hill and into the water and getting ready to transport us out to *Penaeus*, we could see the huge bulky outline of the *Isabel* drawing ever closer. Her worm-eaten timbers and the remnants of her cabin came into view. Then she drew up level with the dock, and she was about five feet off the tip of the T where we all stood in the water, which was over our boots, with rain beating in our faces. At the last moment she passed the dock.

"I hope she keeps on going," said Leon with relief. "Now we got to watch her and see that she don't come back this way or crash into our shrimp boat."

"Well, y'all, watch it," said George, grinning and looking relieved. "If we can be of any help, be sure to let us know. We got to pack out these fish we got in, and get ready for this run of storm mullet that will be coming in after this blow. But if I were you, I'd damn sure tell Sidney Wilson that he ought to do something about his boat. . . ."

The *Isabel* kept on drifting, but the wind began to pick up and she was whisked along with even more force until she ran aground on top of an oyster bar at the head of the bay. There she remained, lying on her side, never to move again. But our problems were just beginning. "Look at those clouds," said Doug, pointing to the sky.

"They're trying to form a twister." Off in the distance we could see long, ugly gray fingers, "witch's fingers," pointing down to earth.

"Let's get the hell off the dock," said Edward nervously. When you see those things forming, you'd better watch out!" The wind started to blow, first with a refreshing coolness that seemed to breathe oxygen back into the stifling air. Then all around us there was a slow swirling, as if all the clouds and puffs in the sky were getting together for one big dance. It was as if the heavens were arranging themselves, getting into position to come roaring down upon the earth, and here we were on our toothpick of a dock that projected out into the water.

The sky became frighteningly black, and as we hurried to the shelter of our most solid building, we could hear the monstrous giant above. The tornado had formed. The screaming winds sounded like a freight train rushing past. A mile away, a neighbor's roof was torn off, whole trees were uprooted and hurled to the ground, power lines came down. A real estate sign was ripped up and whisked down the street and the winds drove the seas hard against our dock.

In just a few minutes the spinoff winds took the farthest portion of our dock, wrenched it up and bent it sideways, giving it a gruesome warped look, as if some giant had reached down from the sky, grabbed one side of the dock's T-section and pried it out of the mud. One of the floating Styrofoam docks was ripped loose from its moorings and pushed across the bay, ending up in the marsh. We could see its outline on the far shore.

Looking at the damage that evening, I was sick. Disastrous as it was, I knew that I had a lot to be thankful for: it could have smashed our home to bits, or demolished our laboratory. Instead of just wrenching up some pilings, that sudden storm-spawned tornado could have taken our dock and rent it to splinters. When the storm would finally quit, we would take *Penaeus,* who fortunately managed to stay at anchor, and tow the floating dock back into place. Somehow we would jet the pilings back down and try to straighten the dock, but it would always bear the scars of the tornado, reminding us that it was ephemeral, subject to the whims of nature.

*121*

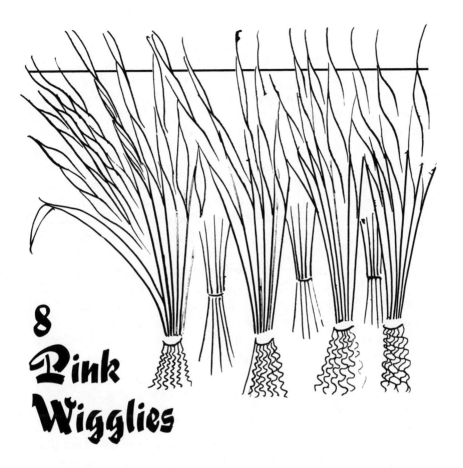

# 8
# Pink
# Wigglies

For three consecutive days it rained, and there wasn't any sign of it stopping. We awoke in darkness and looked out over the angry bay with its black storm clouds still overhead, and nowhere was there the first sign of blue sky or sunlight. The whole house felt wet, our sheets were soggy from the humidity, the air was heavy, and still it rained on. The tropical depression still sat malevolently over the northeastern Gulf of Mexico and was being held there by a weak cold front that was trying to push down from the north. The great forces above us were locked in mortal combat and down below on earth we were getting drenched.

Panacea was practically flooded out. Water was standing foot deep in yards and flooding into houses. The drainage ditches were overflowing, pine and palmetto flatwoods were standing in water, and the swamps were flooded. I drove around Panacea with the rain beating monotonously against my windshield, listening to the even more monotonous wipers whisking back and forth.

I couldn't help but admire it, especially when I drove around the head of the bay and looked over the great green-brown plains of marshland on Fiddler's Point. Nothing is more magnificent than a rainy marsh, especially when a squall brings the driving force of wind and rain sweeping down upon the grasses, blending the sea, land and air into a wondrous gray-violet mist. Lightning crackled and thunder rolled, turning it all into one violent oceanic mist. I looked at the swollen tidal creeks filled with brown swamp water and wondered how the fish and shrimp were faring. Did the marsh offer those tiny frail creatures any protection from the downpour that came upon them with such vengeance? I doubted it. Whatever lived in those creeks, if anything still did, was getting the same punishment that everything else on land was getting. Perhaps they could burrow down into the mud where they could feel only the vibrations of the splattering water. Could they feel the rolling thunder, or perhaps see the dark skies overhead suddenly split by lightning? I didn't know.

Most of the benthic organisms in our bay were tolerant of low salinities, but how much could they take? In the past twenty-four hours alone, more than eleven inches had fallen and almost all of it would end up in the bay. Already there were signs of trouble to come. The sponges that hadn't been ripped off our floating docks during the tornado were turning moldy and the hydroids had all died off. After the water went down and we inspected the damage, the warped boards and broken Styrofoam floats, we hauled up the crab traps.

They were full of dead, bloated blue crabs and the stench was terrible. The crabs had absorbed far more fresh water than their systems could tolerate. Crown conchs had crawled into the trap to feed on rotten bait and also had perished; a gooey, stinking slime oozed out of their moss-covered shells. One trap contained two fresh-water blue-gill sunfish that had obviously been washed out of the Ochlockonee River and found the waters of Dickerson Bay fresh enough to survive in.

I was worried about the impact of rainwater on the bay. I remembered reading a paper by Charles E. Dawson of the Gulf Coast Research Station in Ocean Springs, Mississippi, in which he described a massive amphioxus kill after an unusual storm. Over a two-day period, seven inches of rain had been dumped on the Mississippi delta and salinities plummeted from normal brackish water practically to drinking water.

They found that the entire population of these wormlike little animals had been totally wiped out. During the routine sampling operations they caught more than a hundred dead ones meshed up in the webbing of their shrimp trawls, and healthy amphioxus are never caught in trawls. Normally, amphioxus stay buried in the sand with just the tips of their heads protruding, with which they filter out particles of food from the plankton. Seldom are they found in the gut contents of fish, because when any predator approaches, they dive into the sand and disappear. But after those torrential rains, scavenger catfish were found with their guts gorged with these peculiar wormlike creatures that the scientific community considered so terribly important. Dawson reported that it had taken four months for the amphioxus population to even begin to recover.

That gave me cause to worry. Normally, whenever we wanted amphioxus, we ran out into the Panacea Channel, dropped a dredge overboard, and hauled in all we needed. We had already depleted our amphioxus inventory, having shipped out the last six yesterday. Customers were calling daily, and I was having a hard time putting them off. There was no substitute for living amphioxus.

Many years ago when I was first beginning my specimen business, I would drive from university to university throughout the Northeast, meeting professors who wanted live amphioxus. All they were able to get was prepared microscope slides and pickled specimens from the large biological-supply houses. Most of these biologists had never seen a live one.

It was difficult to admit that I really didn't know what an amphioxus was. At the time I had little background in biology. I grabbed a zoology book and hastily read up on it. I could see the reason for its popularity. It represented an evolutionary link between vertebrate animals—those with backbones, and invertebrates—those without. Running down the back of these wormlike slivers of pink flesh is a notochord, which is the forerunner of a backbone. It has gills like a fish and no brain except for a slight swelling at the anterior end of its nerve cord.

That really didn't help me much in locating them. I looked at whitened preserved specimens. They were elongated slivers of flesh pointed at both ends. I learned that they lived in sand and that they could swim through the substratum in a flash by merely vibrating their bodies. I returned to Florida anxious to look for them.

Finding them was another matter. I roved the sand flats, digging through hard-packed sand and sifting it through screens. I must have shoveled ten tons of sand before I found the first specimen, and I knew that it would be diminishing returns if I continued. Once, when I was digging clams on the bank of the channel, I overturned a shovelful of sand and two amphioxus erupted, danced frantically on the flat surface for a second, and before I could collect my wits, shot down into the sand and out of sight. The movement was so fast that I wasn't sure I'd really seen them.

I dredged up the whole flat and found six. Surely, that was not the way it was done by the commercial collectors who routinely supplied preserved amphioxus to biological-supply houses. I located a commercial collector in Tampa, and learned that it took a bucket dredge to haul them up. Sight unseen I purchased one. It was a peculiar-looking contraption, with two blades for biting sand, designed to bring up a chunk of the bottom. I unpacked it at the post office, hurried down to my boat, dropped it down onto the bottom of the Panacea Channel, and we were in the amphioxus business. We have been ever since that time.

*Branchiostoma floridae*, the species found around Panacea, was a delightful specimen. They lived indefinitely in the same sand that they were dredged up in, and kept themselves well nourished with running sea water. Shipping them alive was no problem at all. They were hardy, almost rubbery in texture because of the strong muscle bands running through their bodies, and they could survive for days, even weeks, in a bag of sea water charged with oxygen. We were able to keep them so well that we often went out, spent a full day stocking several hundred specimens, and lived off the proceeds.

Finally the rain stopped, and we eyed the ominous black clouds that hung over the bay suspiciously. "What do you think, Leon?" I asked. "Can we run out and see if there's any amphioxus?"

"I don't know," he said, shaking his head. "I don't much like the looks of that weather out there. But I'm so damn sick of staying indoors, I'd try it in a hurricane. We ain't got no sea urchins, no amphioxus, no brittlestars or nothing in the tanks. Maybe we can before it gets too bad out there."

We dressed in our rain suits and slickers, bailed the tunnel boat out, and sped down the bay headed for the channel, where we routinely dredged for amphioxus. The seas were calm enough, the rain had flattened them out to a murky mirror finish. But the air was stifling and quiet. Maybe, I thought, if the weather holds up, we'll be able to dredge up some sea urchins and get the embryology people off our backs too.

However, twenty minutes away from the dock, the winds started to blow and the seas began to get rough. Tunnel boats are glorious for running around in calm seas and shallow bays, they can get right up into the head of a creek and draw twelve inches of water; but when it comes to rough water they are miserable things. The motor, which is mounted in the bow, is lifted out of the water, roars mightily, and comes crashing down with a big splash, knocking everyone about, and very little forward progress is made. That lull had been a fool's lull, just enough to build up our optimism and make us think we could get away with it.

"You think we should call this off?" I called to Leon.

"No, let's push on. It isn't but a little ways out to the channel markers. I want to try in the deepest part of the channel where it's about ten feet. Maybe this rain water ain't got down that deep and there's still some amphioxus there."

We proceeded slowly. The waves lifted our little boat up and sent it crashing down. Every few minutes the bow would break free of the water, and the engine would roar insanely as the propeller spun around in air. Leon would turn the throttle down until the propeller caught water again, and then we would proceed forward with waves splashing over the bow.

At last we arrived at channel marker 22, a tall concrete pillar with a black sign on it with luminous white numbers. Across the channel was red marker 25, and it was the bottom area between the two that we intended to dredge for amphioxus. The channel was a hundred feet wide, and the largest amphioxus populations occurred on the sloping side of the eastern bank, right next to the bottom.

Edward cast the shiny stainless-steel bucket dredge over the side and it splashed down into the water. The large coil of yellow rope that it was attached to sped out rapidly, whisking and snaking out into a yellow streak as the boat forged ahead. Just after the last coil disappeared over the transom, the rope came tight against the tow post. Leon slowed the throttle so that the dredge could hit the bottom and take a bite.

When it hit, it bit hard and jerked the boat backward. Leon turned the throttle to almost wide open, and the tunnel boat strained ahead, ripping a foot-wide swath out of the sandy bottom below. We didn't tow the dredge for very long; it only took a moment to fill up with sand. When we were sure it was full, Leon shifted into reverse and started to back up. That was a mistake. The seas broke over the stern, sloshed over the culling platform, and started filling up the boat.

"Hold it!" Edward shouted. "You'll swamp us. Let's run into it."

Leon made a narrow turn and moved back to where we thought the dredge was on the bottom. The yellow rope floated slackly in the water. As we pulled it up hand over hand, I couldn't help wishing we had taken the nice big *Penaeus* out here to cushion us from all this weather. There was a good possibility that there wouldn't be any amphioxus in the dredge, or perhaps, even worse, only one or two in each haul, which would force us to keep on dredging until we filled the order for twelve. I was already drenched through my slickers, and I could feel the cool water trickling down my neck.

At last the gleaming silvery metal broke the surface. The moment of truth was at hand. Would we have empty lifeless sand or would there be amphioxus?

When we hauled the dredge aboard it told us nothing. There was nothing but coarse tannish white sand packed in it. Not a sign of a worm on top, not a sand dollar, brachiopod, or anything but raw sand. It wasn't a good sign. But we emptied the dredge out onto our screen with a loud pop, and all of us squatted down and started spreading it out.

"Here's one!" cried Edward delightedly. "It's a big pink wiggly." There it was, that pinkish white sliver, digging its way through the sand like a fish swimming through water.

"Well, where there's one, there's likely to be more," said Leon, rubbing the salt water from his eyes. "Let's wash them."

"Don't count on it," I said, helping him lower the screen over the side. We began sloshing it up and down and from side to side until all the sand was washed out and only the largest and coarsest particles remained, along with some shell hash and a few pieces of rock and clay. Then we lifted it, plopped it down on the culling platform, and began poking through the residue. There were two amphioxus, a brachiopod, five or six tiny sand dollars, and an angry blood worm that thrust out its long proboscis.

We made another strike on the bank of the channel and brought up none. While I hauled in the heavy dredge, Edward bailed the boat dry, and no sooner did he get the water level out than it flooded in again. The next strike was made right down the middle of the channel, and it had four more amphioxus. It was hard to believe that we had to lift sixty pounds of sand to get one or two amphioxus. It was like panning for gold.

As we dredged between the two channel markers I couldn't help but feel odd about benefiting from this man-made channel. I hated those big ugly dredge boats that came in and sucked up the bottom and gobbled up the sand and spewed it out into the marshlands, smothering the life beneath them, but I had to admit that the channel had provided a habitat for amphioxus and other organisms, It not only gave us access to and from our dock, but brought us a valuable infauna as well. Normally we would have had to run out three times as far to get the same amphioxus.

You couldn't find any on the intertidal sand flats. They preferred deeper water with fast-moving currents and plenty of food. The channel bottom provided all that. Man had traded off an intertidal grass flat for a deep sandy bottom and nature had taken advantage of it. She healed the raw dredged-up bottom and seeded it with new life. I was always amazed and enthralled at nature's healing qualities and how she managed to make the best out of what man messed up.

I was glad I hadn't been living in Panacea when the channel was dredged in for almost two miles over sand mud and grass flats. I had seen operations like that before and they made me sick with man's senseless destruction. I had seen periwinkle snails climbing up the needlerush marsh grasses, trying to escape the rising black slurry of mud and water that was being blasted into the marsh through the dredge pipeline. When the operation was finally finished and the noise and clatter was silenced, the water drained off, leaving packed sand and dead snails behind. Millions of grass shrimp, fiddler crabs, and killifish, and all the grasshoppers, spiders, and varied forms of life that live in marshes were entombed. Mountains of sand stood where there had once been water, creeks, and grass.

When we hauled up the next load of sand and dumped it out on the screen, it was exploding with amphioxus—there must have been a dozen. They writhed and twisted and slithered down into the sand. "Move, pink wigglies!" cried Edward excitedly. That was his favorite name for amphioxus. "Look at them."

I reached in and grabbed one up between my fingers. "We got our quota right here. Now maybe we can get out on those grass flats and see if we can't drag up some *Lytechinus*."

Leon looked at the horizon, thought a bit, and said, "What the hell. Let's go try it. All we can do is drown!"

With one order out of the way, our spirits had brightened, and we raced forward out into the bay, riding over the ground swells. All around us the skies were ominous and seemed to be getting blacker. The wind was beginning to pick up and a light drizzle was starting. The University of Pennsylvania was spurring us on. They had a whole staff of researchers and technicians waiting on our sea urchins.

As we moved out of the end of the channel and headed over the shallow grass flats, I hoped that our luck would hold up. We had succeeded in catching amphioxus when I had been pessimistic about it, and now maybe I would be pleasantly surprised by catching sea urchins.

The short-spined sea urchin, *Lytechinus variegatus*, was another bread-and-butter item. We had come to take them for granted. When an institution sent in a purchase order for them, we merely hopped into the boat, went out to the grass beds, and caught all we needed. They were used in embryology, and sometimes they weren't fertile and we had to switch to the purple-spined sea urchins, *Arbacia punctulata*, that lived on the offshore bottoms. But during the summer and early fall, *Lytechinus* produced enough eggs for our customers' experiments.

All you do is put a number of these spiny round balls upside down in a beaker of water, inject potassium chloride into their mouths, and they start shedding gametes into the water. The males ooze white sperm and the females exude reddish orange ova, and if you suck up the sex cells and squirt them into a dish of sea water, you can watch the dynamic process of fertilization in a couple of hours. You can actually see the sperm wriggling around the outer surface of the egg. The egg begins to spin, and when a sperm cell penetrates, a crystal-clear fertilization membrane springs up; the nuclei separate and before long the egg cleaves in half. The two cells divide into four cells, the four turn into eight, the eight into sixteen, and so on until you have before you a blastula—a mass of cells that are about to form organs and begin differentiating. There are thousands of eggs to watch in a single drop of inseminated reproductive fluid. Every day, in universities all over the world, there is a tremendous amount being learned from the humble sea urchin. If we ever find a key to the mystery of how life began and what life is, it may very well come from studies performed with sea-urchin eggs.

Usually the water was relatively clear over the tide flats and we could look down and see the waving meadows of turtle and manatee grasses growing on the bottom. On an exceptionally clear day during the winter we could even see the round spiny urchins all bunched together, but not today. Even though we were several miles from shore, the water was the color of coffee.

All we could do was run by landmarks, such as the old dead tree off Piny Island, and the last channel marker out in the bay. It was there that we threw the dredge overboard and watched it sink into the murky water and disappear from sight.

Our dredge was a simple apparatus, basically a light steel rectangular frame with coarse webbing. It was designed to pull lightly over the bottom, not dig into it the way the bucket dredge did. As it moved along, it engulfed all the starfish, urchins, sponges, crabs, and slow-moving fish that were in its path. By holding the tow rope and feeling its vibrations, we could usually tell what kind of bottom we had.

It was clear that we were in the right place. The rope jerked slightly, not hard enough for patch rock and not steadily and smoothly enough for pure sand bottom. As we strained ahead, the rain started to fall, and we huddled inside our slicker suits, miserable in the exposed little boat.

We always pulled the dredge for ten minutes, then hauled it up and emptied it, and if the catch was favorable we put the dredge back over the side. The rain was getting worse—it was now a steady downpour, in our eyes, our mouths, running down our necks, but still we pulled on.

Our boat passed through a shoal of floating sea grass, and, trying to forget my misery, I leaned over to examine a clump. I saw a needlefish darting around the surface and thought that was a good sign. Maybe the rain wouldn't have a severe impact out there; after all, it normally was fairly saline water. I tasted the water, but I could hardly detect any salt in it at all. I really became worried. Sea urchins can't tolerate low salinities. There is no such thing as a fresh-water echinoderm. The only way to tell if they survived was to wait out the ten minutes of dragging and see what the dredge brought up.

Today Leon cut it short. "Come on, let's get her up," he said disgustedly, wiping rain out of his eyes. "I doubt they'll be two sea urchins in the whole catch. A man could catch pneumonia out here."

But Leon was wrong. There must have been fifty sea urchins in the net, all of them dead or dying. The stench was terrible. We no sooner lifted the steel frame out of the water and dumped it on deck than the putrid odor of decay almost knocked us down. Many of the urchins had completely lost their spines, some had rotten goo pouring out of their round shells, others were bloated with their mouths distended. It was a horrible sight. There were moldy black boring sponges that had once been bright yellow, but now were covered with a white slime. All the sessile animals that couldn't flee off shore had perished. Gone were hermit crabs that had been there only a week earlier, and not one little green shrimp (*Tozeuma caroliensis*) was found among the rotting grasses and algae. The handsome encrusting orange and black tunicates that grew on the turtle-grass blades were there as mush. There were scallops with their valves agape, showing their rotten dead insides, and there wasn't a single fish in the entire catch. It was the most dead, desolate bottom I had ever seen.

Edward held his nose and shook his head. "Let's get this crap off the boat and get the hell out of here before I throw up!"

We didn't bother to look. We pushed the stuff off into the muddy dark sea, and as we began to cull, the rains started really coming down. It looked as if night had descended. I pulled apart the stacked-up plastic garbage cans and put one over my head, and Edward crouched down in the bow where there was a little shelter.

It was a long trip back to the dock in the blinding rain, even though we were scarcely two miles away. Leon stood at the bow, holding the motor, trying to avoid the roughest waves and keep the boat in the troughs of the swells. The rest of us sat there looking ridiculous with the garbage cans over our heads, but that was the only shelter we had.

I wondered how long this rain would keep up, and how vast would be its destruction. I knew that as soon as it stopped, we would be able to fill our customers' orders for sea urchins by going offshore into deeper water. We might have to dredge at fifty feet, but sooner or later we would find enough live healthy *Lytechinus* to meet our commitments.

The tide flats would recover. Even though the slow-moving or sessile invertebrates that dwelled among the grass had perished and the mobile creatures had dispersed, I knew that sooner or later they would return. The rains wouldn't hurt the turtle grass, and as long as the grass was intact, it would once again be teeming with life. The fresh waters would eventually be flushed out of the bay and in would come the highly saline waters of the Gulf of Mexico. New larvae would come drifting in and settle among the foliage. Mobile fish and invertebrates would return from the deeper waters. It was a process that went on and on forever.

But when the sea grasses were destroyed by man's activities or some quirk of nature, then the whole environment changed and the rich life that dwelled among the grassy bottoms could be destroyed forever. This almost happened once, when these same sea urchins went out of control and increased in such vast numbers that they almost destroyed the grassy habitats.

It happened at Keyton Beach, a tiny speck of a fishing community on a remote marshy shoreline about a hundred miles south of Panacea where there were vast turtle-grass meadows and tide flats. A friend, Steve Moody, had a bait-shrimp boat and every night he would go out on those flats and drag his big roller nets and bring up hordes of bouncing, jumping, kicking pink shrimp and toss them into vats of running sea water. He knew the flats intimately, every swatch, gully, and contour, and where the grass grew densely and where it was sparse.

Suddenly Steve began to notice that there were increasing numbers of sea urchins in his catch. He would haul his net up and find hundreds of the round spiny balls. They were only half grown, about two inches across, but they were increasing day after day. Soon he was catching thousands. They clogged his nets and loaded his boat down.

Then it became obvious what was happening. The urchins had gone out of control. They were reproducing by the millions and they were devouring the grass. Like a horde of army ants, they munched down the rhizome stems and the leaves of the grasses. They scoured the bottoms of algae, and consumed everything in their path.

Soon his nets came up empty of shrimp and packed with nothing but urchins. Gone were the great mats of algae filled with brittlestars and hairy arrow crabs with their long spindly legs. No more were there spiny boxfish that lived among the grass and sea robins, pipefish, and large sea horses. When their grass cover was chewed away they had to disperse. It doesn't matter whether it is a population explosion of sea urchins destroying the cover, or a fleet of bulldozers leveling a pine-oak-hickory forest for a shopping center or a housing subdivision. When the cover is gone, the animal life is gone forever.

Steve became alarmed at what was happening. He called the state's Department of Natural Resources. They went out on his boat. They hauled up urchins by the millions, smashed them, crushed them, stomped them, but it didn't even dent the population. They considered using lime to kill them, and declared it to be the worst disaster since the urchins went out of control in California and devoured the kelp beds and ruined the fishing.

But in the end there wasn't anything they could do. The destructive urchins ran their course, destroying acres and acres of grass beds, and finally they declined. Almost overnight their population collapsed. They started dispersing and dying, leaving scarred, muddy, eroded bottoms that have only now begun to grow back. Perhaps the urchin kill we had just witnessed had thwarted some kind of natural population explosion. I didn't know.

The rain beat down on us as we headed back into the Panacea Channel, and the bay was filled with whitecaps. Sitting gloomily under my garbage can, hearing the rain thumping down on my head, 1 knew that soon I would have to be making calls to my customers and explaining the facts of life. Someday this rain would quit. The seas would run down, and we would be able to go offshore and drag up the urchins.

But in the meantime my customers would have to wait.

# 9

# The Hurricane Fish

One customer simply could not wait. He had to have his shipment of white shrimp, *Penaeus setiferus*. He was experimenting with molting and if we didn't deliver on time, it would set him back weeks.

We looked at the rain pelting the water and reluctantly decided to try. After all, white shrimp are considered to be more tolerant of low salinities than other species of penaeid shrimp. Often they penetrate far up rivers and creeks where the water is almost fresh. The salinity in the bay was still dropping; it was now down to twelve parts per thousand off the dock, and it would be lower in the more shallow areas.

We didn't have far to go: the wetlands across the bay at Fiddler's Point teemed with shrimp at this time of year. And in the fall, north Florida estuaries reached their highest peak of productivity, but I was worried about what impact this sudden deluge of fresh water would have on them.

With the rain still beating in our faces, we bailed out the tunnel boat and sped away from our battered dock toward the gray soggy marshlands. This time we wore diving masks to keep the water out of our eyes. It made us look rather absurd, but it was effective.

We put the net out on the far shore and began pulling it up the tidal creeks, around oxbows, and up toward the shallow head where we usually caught plenty of white shrimp. I tested the water and found that it was down to six parts per thousand, practically drinking water. The salinity is generally lower at the head of a creek, where fresh water blends with salt as it winds down into the bay, but this was obviously too low for shrimp or anything else.

We turned around and headed back out to the bay, watching the water, hoping to see white shrimp jumping as they often did when the otter boards stirred up the mud. But the water looked lifeless, and when we finally hauled in the flaccid, near-empty net, it surpassed even our gloomiest predictions.

At first I thought there was absolutely nothing in the net that came lightly out of the water. Not a pinfish flapped, not a toadfish or catfish grunted or a blue crab moved. Only a handful of dead smelly sponges and some detritus were packed into the net bag. Certainly there wasn't the first sign of a shrimp.

But when we shook out the meager catch on the culling platform we were greeted with a most astonishing sight. There were two peculiar golden pink fish that looked like semiformed creatures that had been aborted. None of us had ever seen anything like them before. They were less than four inches long, but what they lacked in size they made up for in strangeness. Their slippery bodies gleamed in the rainy light, and they had the appearance of creatures that have been living in a cave, with tiny, almost pinpoint eyes. They looked rather like eels, but then again they didn't because they had whiskers like catfish. And one, presumably the male, had a large external copulatory organ.

Hurriedly we placed the strange golden fish in a bucket and our eyes followed their fins, which ran down their bodies in undulating waves. They swam both backward and forward with perfect facility—highly unusual, because most fish cannot swim backward. Only a very few that are adapted for burrowing can.

"You got any idea of what this thing is?" Doug asked me, leaning over the bucket.

"Nothing. No idea at all," I replied excitedly, picking up one and feeling its soft rubbery body. It was a rather unpleasant feeling. It was a little too soft and blubbery for vertebrate tissue.

"I'll tell you what that damn thing is," said Leon, shaking his head in amazement, "That ain't nothing but a hurricane fish! I think the tornado spawned him."

"It could be some kind of goby or blenny," Doug offered doubtfully. "All that fresh water ran that thing out of the ground, you can bet on it. I'd damn sure like to know what it is."

"I'll bet we can get a hell of a price for it from an aquarium," said Leon. "A fish that unusual ought to be worth money!"

"I doubt it," I replied, losing a little of my enthusiasm. "It's too small and obscure. It might make some fish taxonomist excited, but they don't have any money to buy specimens with. Oddball things like this just help boost our reputation in the scientific community."

That night I called Camm Swift, an ichthyologist friend at Florida State University. He often identified our common and rare fish. He really was an expert: he could pick up a live or preserved fish that we had puzzled over, and in one minute say exactly what species it was. Even when we found something we considered rare on the *Penaeus*, like some deep-water black and white striped flounder, it didn't faze him.

But this time he was stumped. He looked in the tank where we had the little fish, gasped with wonder, then pulled his books out of the car and began reading. Finally he shook his head. "I give up. I think I know what it is, but I'm not sure. I think it's a whole new genus and species that's just been described from Mississippi."

Then a strange thing happened. The larger of the strange little fish began to contract and squeeze its body. At first we thought it was going into some sort of death spasm, but it wasn't. It squeezed out a tiny pink baby, and another and another until there were four.

"This is crazy," stammered Camm. "First we find this rare fish and now it gives live birth. It's too much to ask. Do you realize how rare it is to find a viviparous fish? Let me take it back to my lab, where I have my complete library. I'm going to find out what it is, or see if it's an undescribed species."

Later that night he called me. "I found it," he said exuberantly. "It's just as I thought. It's called *Gunterichtys longipenis*. It's related to a cusk eel."

"Why is it called *longipenis?*"

"Because it has a very long penis," he said, laughing "Rumor has it that when Charlie Dawson found it in Ocean Springs, Mississippi, he named it *Gunterichtys* in honor of his major professor, Dr. Gordon Gunter. Gunter didn't like the species name and told Dawson to change it. Dawson said he could have a choice between *longipenis* or *brevipenis* or nothing. So the *Proceedings of the Biological Society of Washington* has a paper entitled '*Gunterichtys longipenis*, a New Genus and Species of Phididoid Fish from the Northeastern Gulf of Mexico.' . . . It's a weird fish. No one knows anything about it. You've made a bigger contribution than you know." His voice was excited. "It's only been found after hurricanes and catastrophic weather conditions like all this rainfall we just had. They found one specimen near the tailings of a dredging operation, and Dawson reported some found by students who were sifting mud. But this is the first hint of how it reproduces—no one had any idea that it was a live bearer."

When I hung up I was feeling good. It was nice to know that some good had come out of that rotten weather.

Several days later the rains finally quit as the cold front grew stronger and the low pressure played itself out over the Gulf of Mexico. The great battle of the heavens was over. Both sides had lost, and the first wonderful bright yellow rays of sun peeked down from behind the clouds. Looking up at the warm golden glow, we knew that up above all that gloom and wetness was the same benevolent sun god that had been there all along. For a while we thought he had forsaken the world and gone off elsewhere, but when the rays reached down to us that afternoon, we knew that we'd been mistaken. We had just lost our faith.

Nothing, nothing ever felt better than those first warm rays bathing our cold clammy skin. We knew the world was reborn again and that somehow everything would work out well. A cooling breeze was coming through and all the mugginess was passing.

Several weeks passed and the south winds started blowing, mixing the highly brackish water in the bays with the salty waters of the ocean. Day after day the salinity climbed slowly from ten parts per thousand to twelve-to-fifteen. It was time to see if we couldn't fill our orders for white shrimp, so one afternoon we loaded up the tunnel boat and headed across the bay to the creeks of Fiddler's Point marshes. The rising salinity came too late for the sponges. Dead sponges were prominently floating on the tannin-stained water that continued to trickle and drain out of the swamps and woodlands.

The sponges were all *Tetilla laminaris*, commonly called "turd sponges" or "porpoise turds" by fishermen. Rotten, swelled with gasses, and bobbing along the surface, they really looked like their unsightly namesakes. Soon they would be assimilated and broken down and would reenter the flow of life As we traveled over the murky water, I looked down at the floating debris that was everywhere—oak leaves and twigs and pieces of bark, leaves from bay trees and sweetbay, even entire dead and decaying sabal-palm leaves. I knew that I was seeing another major aspect of the ocean's nourishment and its dependence upon land for food.

The decaying leaves and other plant materials that flowed out of the swollen swamps, the cypress-stained waters that carried humic acids, tannins, lignins, and proteins, all of this was part of nature's cycles and systems. Soon the oak leaves that floated on the surface would be covered with fungus and slime molds, then would break down and become even richer in protein by nitrogen-fixing bacteria that abounded in the system. While they were decaying, white shrimp, grass shrimp, baby mullet, croakers, and fish so tiny that they can hardly be seen would be picking at and eating them. Even the larger fish would feast upon the swamp vegetation.

The sun was setting behind the pine trees when we put the net out on the opposite shore. In a few hours we would be shrouded in darkness. It was October now and the days were getting shorter. But even in the dimming light we could see an occasional shrimp leaping out of the water in the wake of the boat. It was good to see the bay alive again.

The water around the boat was full of tiny blue crabs swimming along the surface, most of them scarcely an inch long, and I wondered if they'd been flushed out of the marshes and pushed into the bay. It was odd to see so many of them because they usually stayed hidden in the shallows. They danced and whirled around with one tiny claw pressed to their shell, the other outstretched with perfect facility and ease on the surface of the water. I noticed that one crab had an earwig, an upland insect that had been washed out of the swamps with the vegetable matter. The juvenile blue crab held the insect tightly clutched in one claw and with the other claw buffed off other crabs when they tried to come near. I wished I had brought my camera, because it was a good documentation of the association of land and the energy flow of plants and insects to the sea. Right now that was important data.

As we headed up toward the head of the bay where towering pine trees, palm, and oak-scrub hammocks bordered the shoreline, I was filled with anxiety. Because now, just as I was starting to get a scientific foothold in this bay, and just as I was beginning to understand some of these relationships between wetlands, drylands, and waterlands, the woods and fresh-water swamps were doomed to be turned into a subdivision. I knew that all my fighting and arguing couldn't stop it. I lived in a small growth-crazed community that wanted the "better life." Already the trees were crashing down for the new Panacea Shopping Plaza. As the town began to grow, the timber line was thinning out at the head of the bay. Soon there would be housing subdivisions to replace the trees.

It was a gloomy prospect. Dr. Robert J. Livingston of Florida State University had demonstrated that clear-cutting by the paper companies along the shoreline of parts of Apalachicola Bay had an adverse effect on marine communities. When all the trees were sawed down, the ground chopped, ditched, and bedded, the run-off waters contained excessive, harmful amounts of tannins, and humic acid far greater than what a natural system would produce. After intense rains like the one we had just had, the overflow, stained like overly brewed tea, chased away the fish, shrimp, and crabs. Everything in the bay, from bacteria to sharks, had been impacted.

Panacea's problems would be even worse. In a few years, instead of nutrient-rich swamp waters running out into the bay there would be water from drainage ditches carrying pesticides and fertilizers, seepage from septic tanks, and oil from the driveways of family garages. Combine that with the run-off that already existed from the crab houses and it would be a real mess. Contemporary man and nature just didn't mix.

Pelicans and herring gulls, on the other hand, appeared to think that man, especially the commercial fisherman, was the best thing that ever happened. Before we knew it we had five big pelicans flapping their wings and following behind our tunnel boat. They learned to recognize a fishing boat, especially one pulling a net: it meant that food would be forthcoming. Soon five of these rather absurd-looking birds plopped down on the water and began swimming next to us like tame ducks.

Normally one only sees pelicans sitting on pilings or winging their way across the bay, gliding over the surface and then crashing into the water with a mighty splash to snatch up a fish. They have keen eyes, and can spot a minnow from a considerable distance and then pounce. But the days and days of rain had turned the water so dark and turbid that they couldn't see their prey, and they were hungry. So they hung around fishing boats, begging food from hook-and-line fishermen at the dock. There were pelicans everywhere.

They were especially fond of hanging around shrimp boats. In Dickerson Bay when the white shrimp are growing in the marshes, fishermen run the shoreline in their tunnel boats, travel up the creeks, and tow their nets both day and night to get white shrimp. Whenever a shrimper went out into the marshes he was followed by a parade of hungry, squawking, fighting birds, snatching up the discards as fast as he culled them.

Doug was preparing the buckets and Styrofoam boxes for the forthcoming catch. He dipped up the tealike waters and poured them into our plastic garbage cans. We had the boat filled with containers, but that gave us just a shade of apprehension. Perhaps we had taken too many. We had developed a sort of superstition in collecting that generally seemed to be true. If we took too many buckets on a collecting trip (be it walking on tide flats, beachcombing, dredging, diving, or whatever), we never caught enough to fill even one bucket. If we went just a shade unprepared, then we would catch plenty and be looking around for space to put our specimens.

We were towing up the mouth of a large marshy creek that wound its way through the wetlands of Fiddler's Point when Leon became excited. All around us we could see the ghostly forms of shrimp jumping and showering out of the water, leaping into the air

like popcorn and sinking down into the mud again. The waters were literally raining shrimp. It was a beautiful sight. They were swimming on the surface, jumping, leaping and bouncing in a nervous manner, and sinking to the bottom. When the birds saw the splashing, they went insane, snapping and gobbling, but they didn't do very well.

"Goddamn, we're really getting into the shrimp, Jack," Leon said excitedly as he turned the handle on the outboard motor. "We might just load the damn boat down."

"Wonder why there's so many," I said, watching another shoal of white shrimp jumping out of the water. "Maybe the fresh water ran all their predators out of the bay."

"It ain't that," Edward said. "Ain't nobody been messing with them much. It's been raining and bad weather, and no one's been up here with nets scattering them all over the place."

The sun was almost down, with only a few rays of light coming through the piny woods ahead of us. I looked over the stern, back to where my dock was. A year ago there had been only a few lights there, and now there were new buildings and the blazing yellow lights of the Junior Food Store. More trees had been chopped down, and more were coming down every day. It was inevitable that someday the entire shoreline would be a sea of lights pushing back the darkness, the spirits, and the serenity of night.

Doug lit the gasoline lantern as the sun disappeared and darkness descended. Then we could see the eerie banks of the high marshes. The red cedars along the banks, the flowering white high-tide bushes, and the shadows of the sabal palms stood out starkly as silhouettes. Leon turned the boat and we towed the net through the night, around a large oxbow and back down the creek into open water. "All right," he said, "let's get her."

As the net emerged from the water, his voice was filled with delight. "Goddamn. Look at all those shrimp!" Grunting and straining with all our might, the three of us hoisted the net up onto the boat.

Mixed in among the huge mass of decaying sponges, the silvery bodies of fish, seaweed, mud, and debris, were shrimp, shrimp and more shrimp. They were in blurry, dizzying numbers, and when Leon opened the net they rained down onto the deck, jumping shrimp, leaping shrimp, shrimp with orange eyes, jackknifing bodies, and long whiskers. There were big shrimp and little shrimp, all of them lovely and delicious and wonderful. It had been months since we'd really had a decent dish of shrimp. Oh, how I love shrimp!

"There must be close to a hundred pounds in here," he chortled joyfully, and raked through the mass of jumping creatures. "We can get top dollar for them from the fish house right now, and we can damn sure use the money."

"No way. I don't want to sell them. We'll divide them up. Anne and I haven't had fresh shrimp in ages. I'm going to fill my freezer."

"You'll soon get so many shrimp you won't want to look at

another one," he retorted, grinning. "When that first big northerly blows, they're gonna move. I promise you that! The wind's gonna push the tides out and the whole world will go dry and these shrimp are gonna leave these marshes and rivers. There's going to be a hell of a run, and we'll be waiting for them at Mud Cove with *Penaeus*. We ought to make three or four thousand dollars before you know it." He was obviously infected with the dreaded local disease "shrimp fever," similar to the malady gamblers suffer from.

I picked up fifty of the most lively jumping shrimp and placed them in a bucket of water. Our poor customer had just about given up hope and was probably working on white rats by now. The rest of the catch belonged to us. It was here that Gulf Specimen Company paid its real dividends, being out on the bay on a pleasant warm night with calm seas, catching delicious shrimp for dinner and looking at all the creatures.

Little blue crabs were charging off in all directions. There must have been hundreds of them, with their pincers outstretched, snapping and grabbing onto everything. There were baby catfish, both the gaff topsail and the common sea cats with their handsome blue skin, and glittering menhaden. Baby flounder flounced and jumped their flattened brown bodies and stared at us reproachfully with their two eyes on the tops of their heads. A stingray pounded its wings and lashed its poisonous barbed tail. The night air was filled with a grunting, flapping, splattering uproar. Life, life, life—there was so much of it. Puffers gulped air and inflated themselves into little balloons so taut that they looked like they were going to explode.

As we culled, the pelicans were gorging themselves. It didn't seem to matter how much trash fish we raked over; they were making up for lost meals, and they ate until they could barely fly. But soon it was all over for them, since the shrimp were falling off. We headed back to the dock. It was only eight o'clock, but it was dark.

At the floating dock I turned on the powerful floodlight that turned darkness into daylight. Now the light was worthwhile, and by some miracle it had escaped damage from the tornado. We spread out the shrimp and began heading them. That is a long process, where you pinch the heads off between your thumb and forefinger and you look at the mountain of shrimp ahead of you and wonder if you'll ever get through. But with the four of us squatted down, pinching off the heads and throwing them overboard, it didn't take long.

In the water next to the floating dock a veritable feeding frenzy was taking place. Flashes of gold and silver were seen in the dark waters, as fish caught up with the mania of fresh shrimp heads struck and gobbled. There were all kinds of fish down there that hung around the piles of the dock, hordes of pinfish, silver perch, and yellow tails, all with their glittering colors that caught the light overhead. As the shrimp juices permeated the water, the fish came to life and ate and ate. If the ocean were filled with sharks as voracious as pinfish, it would not be a safe place to wet one's toes. These little fish were quick as lightning, alert and ready to attack anything.

I watched another load of shrimp heads go over the side and they barely sank before they were being knocked and jostled and shaken. The more we culled, the more fish appeared, and I could see flashes of larger fish coming in—perhaps it was big speckled trout or redfish. All around the dock I could hear leaping and splashing and jumping. And then a dark form appeared just out of the light and a dorsal fin cleaved the water for a second, and there was that exhilarating feeling of shark. Even though the shark was scarcely three feet long, he was a monster amid the little ones. Probably he was in there hogging his share of shrimp heads, perhaps he managed to snatch up one of the scavengers, but I was glad to see him partaking of the food that we had provided.

I felt good about throwing our shrimp heads back into the sea. If food is taken from the sea, food should be put back into the sea. Unfortunately, that isn't happening to the hundreds of millions of tons of proteins that are being removed all over the world. When the shrimp were headed, we poured them into ten-pound bags and I distributed them among our crew.

What a glory it was that nature could rebound the way she did. Only a few weeks ago this same area had been a biological desert, the empty kingdom of a peculiar-looking slippery-skinned "hurricane fish" that lived deep in the mud, and now it was once again massed over with its enormous diversity. No matter what *nature* threw at Dickerson Bay, sooner or later it would recuperate, flooding us with shrimp, blue crabs, and other good things.

# 10
# Mud Is Good Stuff

Even though the fresh water had devastated the fouling communities and the tornado's spinoff winds had damaged a section of the dock, a vibrant rebirth was taking place. It was now November, and the pilings that had been bare and denuded of life for the past month were starting to blossom once again. The life seemed to spring up around us as we nailed back boards, bolted the Styrofoam floats into place, and jetted down the uprooted pilings. The cool weather blew down from the north, making our labor more pleasant.

The sea-water temperature was slowly dropping. It was now down to seventy-five degrees; a month before, it had been at least ten degrees warmer. No longer did the water feel hot, sticky, and tepid in the bay. As it cooled, schools of mullet moved in to browse on the detritus. We watched porpoises charge into their midst, slapping them with their tails, making them leap into the air, and herding them up for the feast. An alligator took up residence under the dock, a little fellow about four feet long who liked to swim lazily around the pilings and snap up minnows or eat discarded bait. Early in the morning, every morning, a blue heron walked down the entire length of the dock, snatching up little black crabs from the spaces between the planking.

Almost as if by spontaneous generation, the dock was suddenly massed over with *Ecteinascidia turbinata*, the glassy tunicates with transparent bodies tinged with red, giving the zooids a rubylike appearance. For the first time I noticed dense carpets of the fouling anemone, *Aiptasia pallida*, with their dark greenish-brown bodies and whitened tentacles clinging to the Styrofoam, just above the sponges and below the water. There were brand-new plumes of pink and red hydroids erupting from the dock and mats of orange and brown bryozoans. The barnacles seemed to have grown with a vengeance, not only on the wharf pilings but on our boat bottoms.

All the nutrients that had been stirred up and flushed out of the marshlands the month before were probably responsible for the re-birth. Dr. Philip Butler of the Gulf Breeze Laboratories in Pensacola had demonstrated years ago that the waters were much more productive after a storm. For several years he had hung plaques out in the bay and studied the colonization, and found that after there had been a hurricane, oysters, barnacles, bryozoans, and other fouling creatures increased their numbers two- and three-hundredfold. The trillions of barnacle larvae, embryonic forms of bryozoans shifted by the winds and currents, the tadpole larvae of tunicates, the veliger larvae of oysters, and the planulae of hydroids found refuge on our pilings. The dead were replaced with the living. Everything grew and grew until the cold weather, and then life would turn brutal.

As the cold November winds descended on Panacea, the water temperature dropped, and the plankton perished, as it did every year. As winter started, the water was crystal clear and the hungry tentacles of the filter feeders on the dock found lean pickings. Then the tides began to fall, exposing mud flats and pilings that had never been exposed before, and softbodied water-loving creatures felt the burn of the winter winds. The only soft creatures that survived were the ones fortunate enough to live on the bottom of the pilings at the end of the dock, where they were not exposed to the low tides. The barnacles survived the winter with little trouble. When the tide was out and the vicious wind came down upon them, they drew in their feathery legs and remained shut until the water returned. The oysters snapped their valves shut to protect their soft insides from the drying cold, and so did the mussels.

The floating docks dropped down to the base of their pilings and sat squarely on the mud flats with only three or four inches of water beneath them. I had seen winters when the tide went so low that the flats went completely dry and the floating docks, like the boats all over the bay, sat hard aground. When that happened I could see the entire naked bottom, crisscrossed and in some cases deeply slashed by propellers of outboard motorboats that had run aground and dredged their way across the flats. Only yesterday the bay had been filled with water and now it was almost all gone.

Even though I had lived in Panacea for almost a decade it was always a bit of a shock to see Dickerson Bay exposed on the low, low winter tides. It was a good-sized bay, almost two miles long and a half mile wide, and was as handsome as any bay you would find when it was filled with blue water against the brownish green marshes. But when the tides went out as far as the eye could see, as they did in winter, there were only vast acreages of exposed mud. The only water left was in a narrow channel in front of my dock, two hundred feet at the widest, that looked like a small river. The rest was mud.

For just a little while I could walk out on the sea bottom and see what lived there, carrying my bucket and picking up creatures that were normally covered by three feet of water. How strange it was to be walking on boggy lands that only a few hours ago porpoises were swimming over.

I walked out on the tide flats, stepping forward, sinking, lurching forward until I found harder substrate, and stood looking at all that newly exposed mud, mud that was there for the exploring, mud that was there for the taking. Mud is good stuff. We all know it's good stuff when we're kids, but as we grow older our parents and society work on us, they persuade us that we shouldn't have anything to do with mud, and we are eventually beaten down and repressed. Kids know that good things live in mud, like frogs and crawfish and worms. They follow their instincts. But then later they follow the rules and eventually parrot them to their own children.

Every year rivers carry countless tons of mud and silt out into the sea, where it spreads out and builds up the shoreline. Marshes grow in the mud, and shrimp like to bury in it. The process is eternal. Mountains are washed into rivers, wind and frost break rocks into fine particles of dust and silt, and the fine particles move in suspension as part of the water. There is something satisfying about scooping up a good clump of wet mud in your hands, something very basic. There is little doubt in my mind that when life started on this planet, it began in some warm sea on a soft gentle mud bottom.

Standing in the mud, my dock was so strange, almost as if it were undressed with its naked pilings bared to the harsh winds. Those tall pillars now looked ragged with their shaggy beards of hydroids and big bushy growths of oysters and barnacles. All around the bases of the pilings were sponges and conchs and clumps of oysters and other life. Even though incredible gusts of icy air blasted into my face and made my nose run and penetrated through my jacket, I was enthralled with this soft brown wonderland.

I walked out among the pilings, squishing down a few inches through the soft bottom, stepping over oysters. There at eye level were columns of barnacles clicking and bubbling their complaints about being exposed to the cold air. I could see that the hydroids were shriveling and drying up, and I knew that by the time the tide returned they would be dead.

At the very end of the dock, where the Styrofoam floats were, sat *Penaeus*, her hull pressed deep into the mud and leaning slightly. There was some water under her hull, a foot at the most, but her barnacle-covered belly was exposed and ugly. The copper paint was practically worn off. It only showed pretty and red above the waterline where fouling organisms never grew. I knew that we had to haul her out on dry dock soon.

But this was not the time to worry about *Penaeus*. There was too much life around me, too much to look at. I hadn't seen the bay exposed like this in six months and it was fascinating. The floating Styrofoam dock was sitting hard aground, squarely on the mud flats with only a few inches of water around them. And next to them was my favorite fish knife, which had fallen overboard earlier this summer, a little barnacle-covered but still in good shape. There were old reels and rods around the base of the pilings, which fishermen had lost last year, and one of them was in good condition. There were all sorts of bonanzas off the dock, but then again there always were.

I looked at the muddy Coke bottles and some cans sticking up from the mud. That was a never-ending battle with the people we let use our dock: they couldn't resist throwing garbage off it when I wasn't there to raise hell about it. I found it aesthetically offensive, but now xanthid crabs lived in the mouths of the Coke bottles. The mayonnaise jar was encrusted with serpulid worms and slipper limpets and one beer bottle had a large clump of pink coral growing in it. I gratefully put the beer bottle with coral into my collecting bucket. We needed fresh coral that day for an order.

I looked at the pile of rocks we had dumped off the dock several years ago. We bought three dump-truck loads from a limerock quarry and hauled them out to the end of the dock in a wheelbarrow and dumped them over. We hoped to establish a habitat for rock-dwelling creatures and minimize our dredging operations offshore. Then we went offshore and loaded the deck of *Penaeus* down with heavy piles of limestone and carried it back to plant, hoping to keep the sponges, corals, anemones, and other encrusting creatures alive.

Looking at the mud-covered rocks now, I knew that the whole idea was madness. The offshore encrusting organisms that required stable temperature and salinity had long ago perished. The water in Dickerson Bay was turbid most of the year, except for a few brief months during the winter, and the living rocks had come from clearer water where there was no suspended silt.

Nevertheless, fauna and flora did develop on the rocks we had planted off the dock. There were oysters and barnacles, which was to be expected because they grew on any hard surface. But there were also thin orange encrusting sponges growing on the rocks, sponges so thin that I could barely scrape them off with my finger-nails. There were also patches of boring sponges and clumps of finger sponges that had taken over, and among them cheery little sprigs of sea whips, *Leptogorgia virgulata*, grew up. Some of the sea-whip colonies were still in water at the base of the rock pile, and I could see their magnificent fuzzy white polyps expanding from their horny exoskeletons.

I began to lift some of the larger rocks out of the way, marveling at the rich growths of bright green sea grapes that had overtaken their undersides. I saw large brown toadfish hidden in the spaces between the rocks, and heard the loud *pop* of pistol shrimp. When I lifted a particularly large limerock I saw a speckled black and white moray eel slither away. I tried to catch it, but it was too well protected by a fortress of sharp oysters and barnacles.

Beneath some of the rocks were a number of ruby-red cleaning shrimp. The blasting northerly winds had whipped down through the crevices and chilled the handsome little creatures, and they lay there immobile and barely moving their delicate legs while I picked them off. Some had already frozen to death, or died of dehydration when the winds swept through the spaces in and between the rocks, but others began to move feebly when I put them in my collecting bucket. They might have survived until the tide returned.

But for many of the little blennies and gobies and belted sandfish that lived in the rocks, the return of the tide would be of no use. When the sea had been pushed out rapidly by the ferocious north winds, they had been trapped, and now they were dead. I picked the little corpses up with my frozen fingers, and they were a sorrowful sight with their tiny glazed eyes, open mouths, and fins that had dehydrated to a papery texture. The blennies and gobies have no scales, only smooth wet skin when alive, but now they were dry and leathery.

When I saw the handsome belted sandfish lying there on the rocks, I was hopeful that I could revive them in water, but it didn't work. Had I been there an hour earlier I might have been able to save them. They were valuable aquarium specimens, but now their dried scales flaked off in my fingers and blew away with the wind. The wind had dried the algae on the rocks to a fine white powder. There were other signs of death and destruction by the cold; all the long-spined purple sea urchins, *Arbacia punctulata*, had wilted with their spines collapsed around their shells.

I had the thought that perhaps if I hadn't put these rocks there, these creatures would not have died—they would have been living someplace else in deeper water. But then I realized that this was not true. They were there probably because they'd been born there. The habitat was suitable, so the larvae settled and there they stayed.

These limestones made up of creatures that swam in warm tropical seas millions of years ago had brought back life to creatures that lived in the present. For thousands and thousands of years after the sea had retreated, the compacted fossil creatures lay buried beneath Florida sands, lifeless and unused. They were miles from the sea and were sealed off from life. Tree roots had grown down on top of them, and soils had built up over the millennia, until the land had

been stripped off by man's machines and the rocks were dug out to live again.

The surface of the rocks was very much alive, but I expected that. I had always been curious about the rock-boring organisms and how long it took for them to colonize and drill their homes into the limerock. I knew that in the sea it was common for limerock outcrops to be buried by shifting sediment and become lifeless, but I also knew that shifting bottoms could expose new rock and provide new habitats for living creatures.

It had only been two years since the rocks were planted here, and I wondered if that was enough time. I selected a large flat slab of barnacle-covered, algae-grown muddy rock, lifted it above my head, and hurled it down on top of the rock pile like Moses throwing down the Ten Commandments. It crashed down with tremendous impact, breaking into three pieces and exposing the limy white interior. To my delight, the rock was riddled with life. It was almost as alive on the inside as it was on the outside. There were small pink peanut worms, feather-duster worms, lots of little red annelids, and even some boring mussels.

I cracked apart an old barnacle-covered cinder block to see if anything had started boring there, but it was lifeless. The rock borers had to have soft limestone to get started drilling and chemically chewing out their burrows and holes.

But from beneath a small pile of cinder blocks that we had dumped off the dock several years ago I heard a loud bang, which reminded me that we needed a dozen pistol shrimp. I flipped one of the encrusted cinder blocks, and hordes of flattened porcelain crabs ran off in all directions, scurrying among the oysters and wedging their bodies into crevices. There were several young stone crabs that had feebly crawled away or tried to dig down into the mud below. Where there was a pile of rocks, there were usually puddles beneath it, created by stone crabs digging their burrows down into the mud and by toadfish, which also excavated the flats. A whole mini-ecosystem was thus created. In the puddles there would be snapping shrimp, gobies, mud snails, and a variety of other creatures.

I looked at all the rubble on the tide flat. It had a rather unsightly and cluttered appearance, with all those muddy rocks sticking up. But by hauling all those old cinder blocks, broken brick, and old boards down there, we had created a habitat for a large number of organisms.

Once again, man, the manipulator, had changed the ecosystem. A few years ago it was a typical mud flat, just like all the other miles and miles of mud flats around the Gulf. It had its own fauna—razor clams, mud shrimps, polychaete worms, amphipods and so on. But then the rocks rained down upon the surface. The habitat beneath the fifty square feet of mud bottom could no longer support the mud dwellers who needed unobstructed access between the interface of the substrate and the water column to derive their oxygen and food.

One could say that we had "managed" the tide flat. We had managed it for snapping shrimp and toadfish, just as a forester changes a forest by plowing and burning to produce a crop of pine trees. Man likes to manage things. He likes to putter and play and see what he can do to reverse and manipulate nature.

I am no exception. Often I detest the management practices of wildlife officials—when, for example, they have diked thousands of acres of salt-water marshlands, cut them off from the sea, to produce fresh-water marshes where ducks and little else assemble in large numbers. I object to this kind of alteration, yet I can't help doing some of it myself on an infinitesimal scale. Perhaps it's the opportunity it gives me to play God. Let there be snapping shrimp, I say, and down go the rocks and the rubble and before long the rocks grow oysters and beneath them live snapping shrimp and mud snails.

Man manages a habitat for his own benefit. Snapping shrimp were far more beneficial to Gulf Specimen Company than the thin-tubed polychaete worms that once lived beneath the mud. Because they produce sound, scientists and educators find pistol shrimp far more fascinating than the obscure little worms that live in tubes. Often while the teacher is lecturing in the classroom a loud, resounding *pop* can be heard from the salt water aquarium in the back of the room, and everyone knows it's a pistol shrimp. There have been occasions when a pistol shrimp has actually shattered a glass tank by snapping its oversized pincer. The sound comes from a "wrist joint" on the big pincer, and no one really knows why it makes that sound. Some behavioral biologists say that it's a defense mechanism.

I grabbed a heavy flat rock and turned it over, taking great care to keep from getting cut by the razor-sharp oysters that covered its surface, and there beneath the rock, in the puddle, I spotted a pair of pistol shrimp, *Alpheus heterochaelis.* When you first see them they look like miniature Maine lobsters, but have only one big claw instead of two. The shrimp are almost always found in mated pairs beneath a rock, and they are extremely territorial. Seldom do you find more than one pair beneath each rock.

So I spent a considerable amount of time walking over the flats, lifting rocks and grabbing up the pistol shrimp. If the depression beneath the rock was filled with too much water, the shrimp would be impossible to find. No sooner was the rock upturned and exposed to light than the shrimp scurried into the deepest part of the puddle and dug down into the mud.

But in a few minutes I had grabbed up the dozen big pistol shrimp I needed for the University of Virginia, and had six additional to put on assorted general-biology-class orders.

The cold was getting to me, and when I thought about it, my teeth started to chatter. But I decided I would make the cold work for me. It was a good time to collect some highly contractile invertebrates, like *Cerianthus americanus,* the tubed anemone that would be numbed by the cold and easy to catch.

The anemone spun a rather ugly-looking tube of mucus, sand, and mud, and it rode up and down this casing like an elevator. Normally I would walk along the shallows and see them blossoming up with their lovely maroon tentacles, but if I walked too near, they would rocket themselves down to the bottom of their tubes. Only with a great deal of excavation would I be able to catch one.

But on that vicious November morning, with my fingers frozen, I could walk up on the tubed anemone and it didn't move. With a quick thrust of my shovel I was able to block off its escape and pry it easily from the icy mud. And when I broke the ground with my shovel, I could see a multitude of long pink worms, glassy sea cucumbers, ribbon worms, and brittlestars in the mud. I looked at all the mud around my dock—it wasn't mud to me, it was a rich and living substance, the very essence of life itself. All those little pockmarks, casings, burrows and trails, all those little jellylike blobs were living creatures.

I often wished that for just an instant I could see through the mud, that all that opaque brown would suddenly turn crystal clear so I could really see the wealth of life that lay at my feet. Then I would be able to see the millions of clear-skinned mud-dwelling sea squirts, *Bostrichobranchus pilularis*, that revealed their orange digestive organs, and the white-shelled little clams and the gray, thin-skinned sea cucumbers. I would be able to see at a glance the tens of thousands of long-legged brittlestars, *Hemipholis elongata*, sitting seven or eight inches down in the glassy substrate with their long, long spindly arms reaching up from their tiny, almost pinhead bodies. But maybe it was better that I couldn't; perhaps it would be too much of a shock, there would be too much life to comprehend.

There is beauty in all that life. No matter how miserable you feel over your petty little problems, they all seem unimportant and are quickly forgotten when you step out into this ancient world, which has existed for millions and millions of years.

Nature exposes her treasures for only a short time, and when she does, you have to take advantage of it, and get out and collect before the tide returns. I had to get fifty pigmy dog whelks, *Nassarius vibex*, which lived in the mud around the oyster bars.

Getting them in this freezing weather was going to be a problem. A few months before, we could have collected thousands if we had wanted them. Quite often when the crabbers came into our dock and unloaded their catch, there would be dead blue crabs and spider crabs cast off and we could count on seeing hordes of scavenging little brown snails crawling over the bottom, devouring the carrion. They were the sanitation engineers of the tide flats, and during the summer when the water was tepid, they were everywhere. If you wanted to trap them, all you had to do was set some dead fish on the flat, and within a minute or two you would see their proboscises popping up from the mud all around like periscopes.

Then suddenly they would emerge from the sediment in a cloud of mud, bristling forth like army tanks with antennae erect, and start heading directly for the fish. It was no trouble for us to fill orders, we just seeded the flats with dead fish. But this was November, and the water was quite cold. I grabbed a rock and smashed some oysters and spread their juices around, and only then did I see the first sign of life from the puddles I had exposed while looking for the snapping shrimp. One puddle had three or four, but the best catches —of ten and twelve little snails—occurred around the mouths of toadfish burrows. It was a good location for a scavenger, because the toadfish is a voracious predator that charges out of its hole under the rock, grabs even the tough stone crab, chews it up, and messily spits out the shell, and it provides a lot of food for scavengers. Toadfish eat crabs, and they may even feed on pistol shrimp if they aren't bluffed off by the noise.

Predation is the way of life on the mud flats. Everything eats everything else. While the little mud snails had nothing to fear from the toadfish, if a banded tulip shell happened onto the tide flats, they were in grave danger. In warmer weather, when all the snails are crawling about instead of staying buried in mud, the tulip shell and the crown conch rove the flats looking for prey. If a *Nassarius* crosses the mucus trail of its worst enemy, it will violently thrust its foot out and do a back flip to get out of the way.

Nevertheless, I often encountered large snails devouring the pigmy dog whelks. Crown conchs, those handsome bluish gray snails that look like carved chesspieces, are often seen in the upper reaches of the mud flats, among the lowest portions of the salt marshes, patrolling for periwinkle snails. On low tides the periwinkles crawl down to the bottom of the marsh grass and crawl about the mud flats, and when the water seeps out they start their ascent up the grass blades. But often it is too late: a crown conch is upon the periwinkle, smothering it, drilling its long proboscis into its flesh, and in a few minutes only an empty shell is left.

But nothing in the sea is wasted, nothing. As soon as the death scene begins, hermit crabs come out of nowhere and start scampering around the two snails. They come in all sizes and shapes and begin establishing their amusing dominance hierarchies. Big hermits beat off the smaller ones, and the smaller beat off the still smaller ones, until the biggest crab of all stands nearest to the scene of battle, waiting to get the emptied shell. Finally, when the conch does its final contraction and pulls the victim from the shell, the hermit grabs the shell and tries to crawl into it. The crab must never take more than a moment to move out of its older, smaller shell and into the new one. If it exposes its soft abdomen, which is adapted to living in the hard shell of a gastropod for protection, it will be snapped up by passing fish.

I also needed one or two dozen hermit-crab hydroids, *Hydractinia echinata*, which are pink fuzzy coelenterates that live almost exclusively on snail shells that are inhabited by hermits. Zoology professors are especially fond of demonstrating the polymorphism of this hydroid because it has a whole diversity of polyps, some that sting creatures, others that clean, and so on. This was one of the last items on my list, and I was deliberately saving it for last because I knew it would be difficult to get. Only a few hermits could be expected to be covered with the hydroid.

Was it one dozen we needed or two? I consulted my list, checked off the mud snails, periwinkles, and snapping shrimp, and then I noticed that my fingers were bleeding from the oyster and barnacle cuts I had picked up while turning over rocks. The cuts didn't hurt. It was too cold to feel any pain. I tried blowing some warm air on my hands and rubbing them, but it didn't help. I was also getting tired. Trudging through the mud was hard work, especially with the gusting icy winds.

Before me were the great exposed tide flats and the incoming tide. I decided to slog on down to the Rock Landing Dock, where all the crabbers unloaded their catches, and thereby increase my chances of finding the crabs faster. It was less than a half mile away from my dock. I stopped to rest for a moment, and my teeth began to chatter. I knew that if I thought about it, and rested, I would become too miserable to stay out there, so I forged ahead, stepping through the mud. Walking through mud is really an art form. You have to step forward, move your foot to the side and break the suction, thrust your body ahead, and then make the next step. Behind me was a long trail of my footprints dug seven or eight inches down into the mud. In some places there were mudholes and I went down up to my knees, and then I came to sandy areas where the footing was quite firm.

Between my dock and the Rock Landing Dock was a large tide pool that was bound to have some hermit crabs in it. A true connoisseur of seashores would take offense at my calling the big muddy depressions on the tide flat "tide pools." In reality, they were nothing but large puddles left by the tide. Real tide pools are found on rocky shores and usually have colorful assortments of brown kelp, red and green algae, white and pink anemones, blue mussels, small glassy shrimp, and orange starfish crawling around them. You can find these delightful pools along the North Atlantic or the Pacific coast, but you don't find them in Florida.

Nevertheless, you never know what you're going to find in a mud-flat pool left by low tides. Often there are long waving clusters of filamentous algae that are strangely beautiful and look like a woman's hair. And sometimes there are oyster shells overgrown with red and yellow sponges that stand out starkly from the dark brownish black bottom. The real surprises come when you find five or six frilled sea hares, grayish green little blobs of life spewing out long strings of eggs. Sometimes a flounder is stranded in the pools, or a sea robin, and once I found a blazing red and brown scorpion fish with bristling spines.

The big pool on the Dickerson Bay mud flat was only five inches at the deepest part, and when I finally reached it I was not disappointed. It was abounding with life, and there were hermit crabs scuttling about, bluffing each other and moving quite vigorously even though the water was practically freezing cold. I immediately spotted six hermits wearing the bright pinkish orange fuzz. They stood out starkly against their background, and with my cut numbed fingers I picked them up and put them in my bucket.

The big tide pool was also filled with larval shrimp. Everywhere I looked I noted a frantic stirring within the puddle. I bent down to examine it closer and was amazed to see that even in the film of water covering the flat, at the very edge of the pool were millions, literally millions of swarming larval shrimp. So tiny were they that when I scooped up fifty or so in the palm of my hand, I could just see their little black eyes, and their bodies were almost completely transparent. The only thing that gave me any clue to their being there in the first place was the little puffs of mud they made as they scattered off in all directions. Most of the little wraiths avoided the deeper portions of the pool. They clustered around the very edges in just a fraction of an inch of water, just enough to give the top of the mud a wet sheet. There they were protected from some titanic inch-long fish that couldn't make it into the shallows, protected until the water seeped up on the tide flat, inundating the puddle, and the predators came in. Then the little shrimp would move up with the incoming water, hugging the very edge that lapped up on the shore and spread out through the protective jungle of marsh grasses.

I looked up and all over the mud flats were pools and puddles, some of them made by flounder and stingrays that had dug down into the mud, and all of them filled with these nearly microscopic shrimp. I had no idea what kind of shrimp they were—perhaps larval grass shrimp, perhaps penaeid shrimp that had just arrived from the open ocean, ready to start growing in the marshes.

As the tide came in, l actually saw the little shrimp following the sheath. The water moved forward, filling the burrows of the mud shrimp, seeping down into the ground, covering the flat, advancing an inch at a time, and a vanguard of shrimp followed. I knew that in a short time the tide would be in, and unless I hurried I might not find the rest of my hermit-crab hydroids.

But I really didn't have anything to worry about. As the water began to cover the flats and reclaim the land, the dwarf hermit crabs, *Pagurus longicarpus,* were starting to stir around. I found more than a dozen trapped in the tide pools, flaunting their fuzzy pink hydroid-covered shells. The pink polyps didn't discriminate—the bright pink peachlike fuzz covered a variety of crab-worn snail shells, *Cantharis,* moon snails, tiny crown conchs, and even periwinkles. The eye-catching pink against the mud made for a startling and beautiful contrast.

The wind had stopped blowing, and since there was no longer any force to keep the water pushed off the tide flats, it began to rise; I watched it begin to cover my boots. Out in the bay, three vast mud flats were beginning to shrink as blue water covered them. The sea had exposed her treasures to me for a little while and now the sea was taking them back.

# 11

# Of
# Dunes
# and
# Electric Rays

The north winds continued to blow and all over the coast the marshlands and rivers drained until they were almost empty. Only a thin trickle of water was left in the deepest channels in the shallow bays at low tide. Everywhere you went along the coast there would be mud flats as far as the eye could see. Standing on the beach, you could gaze out over the miles and miles of tide flats and see barely a skim of blue water far out on the horizon.

When the sea emptied out of the bays, the shrimp that had been living in their protective mud bottoms in the marsh creeks began to move. Shrimp by the billions began crawling out of the mud, all pointing in the same direction, with whiskers touching tails, and they began to run. They swam along like a cloud of locusts, covering the bottom, moving with the falling tides heading in the same direction, oriented by some unknown guiding force until they reached the deeper waters of the Gulf and congregated along the outer sandy beaches.

It was a delightful morning aboard the *Penaeus*. I was thoroughly enjoying it as we dragged our shrimp net along the beach at Alligator Point. The tide had come back in and Leon was dragging the net as close to shore as he could without running aground. We could see the open sandy beaches, those magnificent windswept dunes, contorted slash pines and scrub oaks in every detail, and they were splendid. The sea oats waved their golden tassels in the breeze and sand blew down the beach, blasting the rugged live oaks, the round needly dark green bushes of rosemary, and the magnificent beach goldenrods. There was a haunting beauty about the dunes, with all those wind-tortured tenacious bushes and trees that clung to life, withstood the salt spray and the hurricane winds, extending their roots down through the poorest of poor soil.

As I sat on the narrow bow, bundled up in my jacket and sweaters, I marveled over the almost blinding white mountains of sand that rose up sharply from the sea. The big orange sun rose up over the water, coloring the sea oats with a fiery glow. The sky was intensely blue, the air crisp and cold, and here I was aboard my very own shrimp boat, about to harvest the treasures of the sea. It was an exhilarating feeling.

Leon was a good skipper. He knew how to shrimp. He could navigate *Penaeus* so close to shore and drag in such shallow water that the otter doors would actually rise out of the water where the surf washed up on the beach. This took a considerable amount of skill because it would be easy to flip over the doors and tangle the nets. The sixty-foot trawl net was spread apart by the two otter doors which were being towed by two hundred feet of cable and the powerful *Penaeus*. These otter doors had to be precisely angled so they would spread the net apart and keep it gliding like a gigantic kite. The trawl that moved at four miles an hour engulfed anything in its path. However, white shrimp and the valuable electric rays, *Narcine brasiliensis*, that we sought, hugged the beach. The closer we could drag the intertidal zone, the more we would catch. Just before we approached the long sandbar that extended perpendicularly from the beach, Leon would swing out and the door that was churning up the beach would sink out of sight as we headed for deeper waters.

"I don't know why it is," Leon said, "but them electric rays will come to this beach every time there's a full moon this time of year and bury up. Some of them are having babies now, so we'll probably catch a dozen little ones. If we dragged even twenty yards offshore, we wouldn't catch near as many. It gets shallow in places, but you can always get in here and shrimp on a northerly if you know what you're doing," he said knowingly. "All the wind is knocked off before it gets downshore."

We were dragging in a wind shadow, and the long peninsula of Alligator Point protected us from the choppy rough sea offshore. "I don't like to come in here on a southwesterly," Leon added. "If you ever break down and wash up on the beach, you can kiss your boat goodbye. It will beat all to pieces."

"Do you think we'll catch any shrimp this close in?" Doug asked him, watching the dunes and the cottages on the beach. We could see footprints where people had walked the beach, we were in so close.

"Hell yes, we'll catch shrimp up here," said Leon, turning the wheel slightly and watching the doors pull along. "After a cold snap like this those shrimp will come to the beach to feed on all that seaweed and mud that's stirred up. That's why they call it Mud Cove right where we're at. Didn't you see them shrimp jumping when we started this pass? We ought to load the damn boat down with them. And, Jack, you don't have to worry, we'll catch all the electric rays we need too. I'll put money on that."

I hoped he was right. We needed twenty electric rays; the orders had been on the books for weeks but we had been unable to fill them. Ever since I began supplying laboratories with marine life I had calls for these odd-looking frying-pan-shaped fish that carried their own batteries and jolted anything that bothered them. I was intrigued by them, which had begun years ago on the beaches of Al-

ligator Point, where I received a firsthand experience of their power.
It was during the summer and the tides were high. I was trying
to collect a number of the burrowing sea anemones, *Bunodactis
stelloides*, which live along the sandy beaches of the Atlantic and Gulf
coasts. It was a rush order and there was no time to wait for low tides,
so I put on my mask, fins, and snorkel, and waded into the surf. But
the water was too turbid, so I gave up and started hunting for the
anemones by braille. I knew I was in the right location, so I moved
my fingers over the bottom like some sort of predatory animal until
felt a soft round blob pulling down into the sand. I would grab it
with one hand, dig it out with the other, and produce a long gray
anemone that first looked like a gray stalk of celery and then rapidly
contracted down into a little gray onion.

All it took was the slightest touch and they whipped in their
stubbly gray tentacles and shot down out of reach. My fingers felt
the contours in the sand, moved over the broken cockleshells and
around the soggy seaweed. I felt the prickly round shape of sand
dollars, and then the soft round anemone.

I was working on an order for fifty, and had already collected
nearly thirty when suddenly I felt a jolt run through me. It shook my
whole body, and I sat there dumbly in shock. I had no idea what had
happened. I was in a daze. Blinking stupidly, I put my hand back
down, and I was promptly jolted again. This time I put on my mask
and peered down through the murky water, and there was a large
electric ray. Even though I had touched it twice, it hadn't budged,
and it acted as if it couldn't care less. It was quite ready to deliver a
third jolt if I were so inclined. This was no lowly stingray who
whipped its venomous tail into some unsuspecting victim and fled
like a coward. This fish was completely aware of its power, and it had
a certain majestic aura about it.

When I got over the shock, and my mind could finally put electricity and water into perspective, I watched the ray in utter fascination. Its soft, light gray body was covered with peculiar markings,
some triangular, others semicircular, that looked like hieroglyphics.
It looked as if some ancient god had inscribed them with an

ominous message that if deciphered would unlock the secrets of the powers of the universe. I prodded the ray with my rubber flipper, and when it saw that its shock was having no effect, it flipped its tail, lifted off the bottom and swam five or six feet away, moving very slowly, and then settled down on the bottom. And then it flipped its tail to and fro and in an instant disappeared into the sand, with only its eyes peering up.

Years later I learned more about the rays, and how they utilized their electric powers in nature. It wasn't only for defense. They used their masses of electric tissue to stun their prey. One day when I was working on a shrimp boat I dissected a number of electric rays and found that all their stomachs were filled with long sand-dwelling polychaete worms. They were the longest worms I had ever seen. In fact, they were completely intact, down to the last anatomical structure. There wasn't a broken parapodium or piece of damaged tissue anywhere. Almost all fish love to eat worms, and they do a messy job of it, tearing and shredding and mutilating the worms. But not the electric ray. I could envision one landing on a worm that had just its head sticking out of its burrow. I could see the ray tensing its muscles, and the electrical energy flowing from its body while it grabbed the stunned annelid and pulled it out like a limp piece of spaghetti. Even though they only generated twenty volts, their power was greatly magnified in the salt water.

It was more than a decade ago when I first came to Alligator Point aboard a large shrimp trawler, working as a deck hand, and saw the nets come up gorged with life and full of electric rays. We had traveled all night from Apalachicola and when I awoke in the morning we were anchored off this magnificent shoreline with its majestic bluffs covered with golden sea oats and dark green scrubby vegetation that stood out starkly in the morning sun. I looked at the wind-twisted pines growing up through the shifting sands, and I felt like Columbus discovering a new world with awesomely beautiful wilderness. There were a couple of beach houses along the shore, and they almost seemed a welcome sight because the area looked a little too wild and strange.

## THE LIVING DOCK

Twelve years later, portions of Alligator Point still looked that way, but only a few acres here and there. Much of it was now an endless clutter of houses, a veritable slum by the sea. As *Penaeus* moved down the beach, dragging her net, we passed row after row of houses built on top of those once pristine dunes. There were still other parts that were unspoiled, but they all had for-sale signs sticking up above the waving sea oats, the creeping beach morning glories, and the thick cover of yaupon bushes with their eye-searing bright red berries. The bulldozers hadn't shoved down all the dunes and torn up all the dense stands of dwarfed sand live oaks, but they had done their share. I supposed that I should be grateful for the time left on this planet to enjoy the few remaining stretches of unspoiled wilderness. In a few years I wouldn't see any of those starkly beautiful dunes. It had taken thousands of years for those dunes to build up, and it only took such a little while for them to be destroyed.

As I saw the houses built down near the water's edge, I thought, How foolish you people are to live on the sea and to tempt her so and risk all you have. Because while the water was calm this morning, with only gentle waves lapping on the shore, I had seen this same shoreline a raging frothing battery of waves pounding at the dunes.

The dunes could take it, but the houses and other man-made structures could not. The dunes were practically alive, shifting, falling, moving bodies of sand that washed out to sea, shrinking now and building up later. They were a dynamic, moving process, and anyone who built on the dunes was in trouble.

As *Penaeus* moved downshore, dragging her nets, we went past older houses, and their trouble was plain to see. Seawalls and sandbags and rocks were piled up against the front of the houses, and still the sea was washing out the sand from under them. There was no building on a shifting, moving, living beach that was gradually eroding into the sea. The dunes mocked the engineers and the builders. It mattered not what they built, sooner or later the sea would get her sand. Sand washed out from the roots of trees and they crashed down into the sea. From beneath the seawalls and the pads of houses, the sands would shift out and be drawn away. In no time the sea would come in, sending her waves smashing, down on those structures, which would crumble and fall. And those foolish people, living in their houses, closing off the seashore from us all, would be so very sorry.

My thoughts were interrupted when Leon said abruptly, "Okay . . . let's get her. Now I'll show you what shrimp look like!"

He slowed the motor down to a crawl and tried to thrust the winch into gear. He stepped on the clutch and shoved, but it didn't want to go. "Damn this power take-off—it froze up again," he replied to my raised eyebrows. "We just greased it yesterday, but I think it's about shot, and there ain't much we can do about it. It's just plain old wore out. We'll have to replace the bearing if we want to keep on using it. It shouldn't cost too much."

But finally the winch went into gear and the engine changed to its usual noisy grumbling clatter, and the winch drum began to turn, slowly wrapping the hundreds of feet of cable around the rusty steel spool. The winding winch half-pulled the boat backward, and the engine labored steadily in the background.

The net was coming in and the angle of the cables became sharper between the boat and the sea bottom. You never knew what the net would bring up. There was no guessing or predicting, only hoping, especially on the first strike. Maybe there would be the electric rays we needed, maybe a big shark or a sea turtle or a net full of shrimp. That first tow is the most anxious of all. After it, the second strike is much better as far as waiting goes, because you have some idea of what kind of sea life has moved into shore.

At last the heavy doors that weighed the net down and made it move through the water spread apart like a kite, broke from the sea with a splash, and were pulled up to the davit arm that hung from the side of *Penaeus*. Leon quickly shut off the winch and started racing the boat forward. The webbing became taut and stretched out, and behind the boat we could see a large ball of life crammed into the bottom of the net. All shrimpers washed their nets down to clean the mud off the shrimp, but we towed it for a shorter period so that we wouldn't damage the life any more than we had to.

With a complicated procedure of running ropes, pulleys, and block and tackle, Leon started lifting the heavy net out of the water. Gorged with life, it slowly emerged dripping from the sea. There were hog chokers and cusk eels, and tonguefish and croakers all gilled in the webbing. And through the crisscrossed cord we could see the white bellies of butterfly rays pressed against the black webbing. But mostly there were shrimp.

"God almighty damn," cried Leon. "We got the shrimp this time. Come on, Edward, open that son of a bitch up! I can't stand it."

The net hung suspended from the mast, shaped like a giant teardrop, and Edward struggled with the complicated knot that kept the bag fastened. Water showered out of the webbing onto the deck, compressing the creatures even further. He snatched the knot loose finally, the net opened, and all the marine life disgorged onto the deck with a great squish.

Out came hordes of croakers and pinfish, catfish, sand eels writhing and wriggling, and flounder. There were stingrays and butterfly rays pounding their wings, blue crabs racing about with their pincers raised, snapping at everything, and jumping and jack-knifing all over the deck were thousands and thousands of white shrimp. And there, on top of the pile, were no less than ten large electric rays, writhing to and fro, exhausted from having expended all their energy. There were also a number of baby electric rays that looked as if they had been newly born, perhaps aborted in the net.

Leon, Edward, Doug, and I were in there busily grabbing them up, pulling them by their tails out of the pile of struggling life and placing them in buckets of sea water. Electric rays can't shock you when you grab them by their frying-pan-handle tails because their two kidney-shaped electric organs are in the anterior parts of their bodies. But they tried. When we picked them up, they bent into an arch, trying to touch our hands with their heads. One succeeded in touching Doug, who let out a yelp and dropped it.

When we had all the rays picked out and put away, Leon said

excitedly, "Come on, we can be going through this stuff while we're towing. Let's get the net out. We're tearing up a shrimp's ass!"

There was life on the deck, life that couldn't live out of water much longer than we could live under water. We raked through the plies of fish and shrimp, looking for still more rays. If they weren't found very shortly, they would suffocate. We jerked out small gray bonnethead sharks and threw them into Styrofoam boxes, gathered up sea cucumbers, and stopped to pick out flame-streaked box crabs. As we raked through the avalanche of life, we found four more electric rays, but the pile was so large there was little hope of finding all of them in time to save them.

Four mated pairs of horseshoe crabs were buried under the mass of life, and Leon jerked them out by their tails and held them up while they clawed and scrabbled at the air, beating their gill books in a futile effort to escape. "Horny little bastards," he said, laughing. "Here they been drug in the net and beat up, and they still won't come undone." He tossed them aside and they started lurching over the deck, thrusting themselves along with ski-polelike rear legs, making a clanking noisy sound.

We didn't give them any further attention, because there were perishable creatures still in the bottom of the pile that had to be rescued. Horseshoe crabs are practically indestructible, and can live for days out of water.

We dived in and culled. It didn't take long to go through the big catch, with four people each raking out a pile of their own, spreading it out, removing all the shrimp, picking out the starfish, sea cucumbers, blood clams, and other prize specimens. Periodically when the piles of discarded croakers, catfish, grunts, pinfish, hog chokers, butterfly rays, and other creatures got too thick, Edward would go around and shovel the pile over the side. I encouraged him to do it often, because the sooner the fish were returned to the sea, the better their chances of survival.

But merely hitting the water was no guarantee of survival. Behind the boat, sea gulls appeared from all over, squawking and screaming. How did they know, and how did so many sea gulls

appear at one time so suddenly? What signal clued them in? Did one see another dipping miles away and start toward the boat, and did another sea gull see the second change direction and so on? Pelicans also dropped down from the sky and began gorging themselves on the fish almost as fast as they were tossed over the side. Two big gray pelicans followed behind *Penaeus* like big domestic ducks on a pond, bobbing their big pouches down into the water and gulping them full of fish. Usually I felt bad about killing all those fish. I would look at their glazed lifeless eyes, or see them gasp away the last of their life, and I felt lousy when I stopped to think about it. Shrimp boats are so wasteful, they kill so much to get so little. If only there were a way to design a net that would catch nothing but shrimp, or nothing but electric rays, and let everything else go. It was true that *Penaeus* destroyed only the tiniest fraction of the rich estuarine life compared to the thousands of huge double-rigged steel-hulled boats with their massive nets that pulled up hundreds of tons of benthic creatures to harvest seafood. But that didn't justify *Penaeus*.

Sometimes during the summer when the hot weather suffocated the fish almost as soon as they were dumped from the nets, I really felt wretched and often thought of getting rid of *Penaeus* and getting out of the whole business entirely. But not that day, not when I saw the pelicans getting all they needed. Many had grown thin and tame since the rains of several months before.

That was a bad situation for pelicans. Years ago, before the human population swelled and vast acreages of farmlands were sprayed with pesticides that drained into rivers, a few lean months probably didn't have much impact on the pelican population. But now, with poisoned waters washing off the various agricultural fields, the pelicans were eating fish that had pesticides in their tissues and were accumulating them in their fat.

Spraying of DDT was so prevalent in the 1950s that the pelican population on the Louisiana coast was exterminated. A massive effort was made by conservationists to reestablish the bird there, and for a while it looked promising. They were nesting and reproducing and the population was beginning to grow. Then, during a recent series of floods on the Mississippi River, nearly eighty percent of the Louisiana pelicans perished in a two-month period, and the pesticide eldrin was found in their tissues. There was considerable debate as to how it happened. Some believed that the birds might have encountered a sudden lethal dose from some unknown source. Others suspected that the pesticide was released into the birds' bloodstreams when they drew on their fat reserves because their normal food supply was interrupted by the muddy opaque water.

The pelicans had to hustle to keep up with the competition from the hordes of gulls that screeched, fluttered, and snatched the food away from the big birds whenever they had a chance. The gulls had excellent vision. They could spot a fraud when one went their way. If a big white cannonball jellyfish was pushed overboard, or a piece of sea pork, they didn't go near it. Nor did they try to come in and grab up the clumps of brown shaggy bryozoans or go after the spider crabs. Edward liked to feed the birds. He would stop culling and hurl a croaker high into the air amidst the hovering, fluttering gulls and watch them swoop after it and grab it in mid-air. Constantly beating their wings, they maintained their distance from the boat, but didn't get very far away either. He would toss a fish directly to the pelicans behind the boat, but rarely did they get it before a gull had it in its beak and was flying off, with its fellows chasing it.

Leon held the wheel while everyone squatted over the diminishing pile, raking out the shrimp. Four wire baskets were practically heaped to the top with the gleaming white bodies of the crustaceans, and their emerald tails shone brightly in the cold morning sun. Every now and then Leon would leave his post, go over to the basket, reach both hands in and lift out the shrimp. "I just love the feel of them," he said, his eyes dreamy. "Just look at them, they're pure gold. There ain't a better feeling in the world."

If he caught one of us missing a shrimp, he would throw a fit, especially at me because I wasn't as attentive as my workers. "Damn it, Jack, you throwing all the money overboard. Look at that big shrimp you just missed, and there's another one. If this keeps up, we're going to have to recull every one of yours. And you too, Doug . . . I see some shrimp *you* missed. Goddamn," he said, shaking his head as if we were all total nincompoops, "I don't know what I'm going to do with you boys.

"Here I took you to raise . . . sent you to school . . . and you eat the books like a goat! Lord. . . . Lord . . . "

This got us laughing, and we culled more carefully. We raked into the pile, spread the catch out, flattening it so we could see what was there, grabbed the white shrimp with both hands and tossed them into a wire basket. Then we would take the cullboard and rake the trash fish behind us and reach for a new unculled pile.

"All right," he said, "let's get these shrimp down below before they spoil. Come on, Doug, Edward, open the hatch." And they lifted the hatch off, lowered the baskets of shrimp, and dumped them on ice and shoveled ice over them so they would stay firm and fresh. Before he put them away, Leon washed them down so that they would gleam better, and then we went back to culling.

I had to agree, they were a pretty sight. There was already close to three hundred pounds aboard, and there was probably another fifty left. "What are shrimp bringing these days?" I asked.

"One dollar a pound and not a nickel less," said Leon emphatically. "If the fish house don't want to pay that for them, I can damn sure ice these mothers up and peddle them from the truck for a dollar-fifty. It don't matter to me . . . I got customers all up and down this coast who's been nagging me for shrimp."

But we were rich for only a little while, because on the very next strike the net came up almost empty. It happens that way sometimes, just when you're starting to make a killing the shrimp will vanish like fairy gold. When the net came up it was full of croakers, and there were two electric rays, catfish, and very little else.

"The tide's turned," said Leon, raking his booted foot through the pile of creatures on deck. "There's not ten pounds of shrimp in this mess. We'd better get offshore while we still can and drag up what sea urchins we can and get our asses back to the dock. It ain't gonna be no picnic out there if we leave now."

We started heading due south, out into the Gulf of Mexico, leaving the dunes behind us, and soon we were away from the wind shadow and the seas began to get rough. He was right. It was no picnic, but when we hooked up the dredge and dropped it overboard, it came up crammed full of live, healthy sea urchins that were full of eggs. We were back in the urchin business.

By the time we were finished, the seas were getting really rough. *Penaeus* pitched and rolled as the seas rose and fell, and the winds began to increase. Waves splashed down over the bow, and even though we huddled around the little shelter, there was no real protection, and we were drenched to the skin. It was dark by the time we made it back into the Panacea Channel.

Off in the distance I could see our dock all lit up, the string of lights that stretched out all over the water ending in an illuminated T. And in the midst of it all was my big blinding squid light that illuminated the water. Anne and Mary Ellen and her husband were waiting down on the dock for our return. The blazing lights grew closer and we could see the outlines of people. Seeing the dock at the end of a trip was always a good sight.

# 12 The Winter Sea

I had on two sweaters, two pairs of pants, a woolen cap, and long underwear, and even then I felt frozen as I climbed down onto the floating dock and turned on the huge floodlight that banished the darkness. Even with that giant globe burning down into the water, I felt the icy, wintry desolation, hundreds of feet from shore, with the clammy wet cold penetrating down into my bones.

Usually when the light was first switched on fish would be startled and would jump and splash, and stay frozen to the bottom, but that night there were no fish around. The water was too cold, but it was crystal clear, and I could see all the way to the bottom. Strangely enough, because the water was so transparent, I could even see the bottom of the floating Styrofoam blocks. They looked as if they were suspended in mid-air, with their tufted brushy hydroids and lumps of sponges that were flowering with tiny pink

*179*

anemones. With just a little luck, there would be *Nereis limbata,* the pink bristleworms living in the canals of those sponges. If I could just bring myself to reach into that freezing water, I might be able to fill the order for the hundred that would be needed tomorrow.

I grabbed up a hunk of sponge and tore it loose from the Styrofoam, and spread it out on the floating dock. There were a few xanthid crabs in it, looking very cold and barely moving, and lots of little pink anemones that had contracted, but not even one worm. Maybe if I had looked hard, and torn off every bit of sponge and the remaining clumps of dead and dying winter hydroids, I might then have located a few dozen worms, but never fifty. It meant that tomorrow we would have to travel offshore in our tunnel boat and scrape the buoys to fill the order.

The silence of the cold dark night was broken when I heard a motor start up, a whirling sound, and then a sound of gushing water. It was the pump down in the bilge of *Penaeus,* pumping out her leaking hull. Water seeped in between her boards where the caulking had rotted out, it rose in the bilge until it tripped the float switch, and then the pump sucked the oily water out and spewed it into the sea. I knew that she had to be pulled soon. Tomorrow we would have to use the tunnel boat to fill our order for bristleworms because a few days ago a piece of trash got hung up under the float switch and *Penaeus* almost sank. Water rose in the floorboards and seeped into the starter and ruined it. We tried to buy another starter, but no one had that particular model in stock, so *Penaeus* was incapacitated until one was shipped down from Michigan. That is the way of boats, always breaking down, always giving trouble. A dock that just sits there day after day is so much better—you don't have to fool with repairs, or fueling up, or working with nets to get marine life. The marine life comes to you, if you can wait long enough.

The tide was falling now and the water was draining out of the creeks, flushing out the wind-dried muds and algae from the cold dormant marshes. Where there was water now, there would be chilly mud by morning. Although there were hardly any fish in the bay, the waters were by no means barren of life. Whenever a light shines on the water there will be life. Periodically long pink ribbon worms were swept along in the current, doing loop-de-loops, writhing and twisting themselves into knots and moving along out of the range of the light. Probably they were spawning, dancing in the freezing water and spewing forth sperm and eggs. Had these long flattened pink ribbon worms that looked like living shoelaces been swimming a few months earlier, or a few months hence, they would have been immediately gobbled up by hungry predators, but now it was too cold and the worms had the water to themselves. Down in the water I could see my baited crab traps and there were only a couple of blue crabs in them, and even then the bait was hardly touched. In the summertime the bait would have been devoured in a matter of hours by hordes of hungry little pinfish and yellow tails that swam in and out of the wire mesh as if it weren't there. Those little fish would have made quick work of the worms that were gliding about so gracefully. So would the blue crabs, but all the blue crabs were nestled down in the mud, sleeping away the cold, waiting until the warm Gulf waters came into the bay so they could stir around.

It occurred to me that if I had to describe the winter fauna of Dickerson Bay, I would have to speak about the species that were absent rather than the ones present. There were no small sharks, no rays, no sea robins or cowfish. You didn't see the big gray isopods, *Ligyda olfersii*, otherwise known as "sea roaches," or the grapsid crabs scurrying around the pilings. They were hidden down in the crevices and cracks in the pilings and wouldn't be seen until spring. What happened to the thousands of little water striders? A few months ago they were dancing all over the surface, whirling about in dense masses, and then suddenly they vanished. We saw no mass migration of the feathery little insects; if they hibernated for the winter we had no idea where they went.

The great masses of jellylike purple tunicates, *Ecteinascidia turbinata*, that had suddenly massed over everything else on our floating docks and oyster strings had died. I had seen them turn moldy and fall to the bottom when the water started getting cold. I had seen the *Eudendrium* hydroids perish, watched their polyps fall off and saw just their bare branches remain behind. The dead branches were still attached, with one or two forlorn skeleton shrimp hanging on, bobbing elongated bodies feebly.

Two more *Cerebratulus lacteus*, the ribbon worms, swept past the floating docks on their way out into the bay, undulating their pink bodies amidst a dozen cross jellies that were pulsating along in the night traffic. Now there was a peculiar animal, and one that could be said to be part of an exclusively winter fauna.

*Nemopsis bachei*, the glassy-clear hydromedusae that consistently appear in December, are scarcely a half-inch in diameter, yet they move through the water with endless vigor. Four lobes of thin branching tentacles protrude from the margins of their transparent bells, looking like the most delicate of threads trailing behind the moving animal. We called them "cross jellies" because the two U-shaped radial canals and folded gonads that cross in their bell-shaped umbrellas were the only visible part of them. On a clear day, if you stood on the floating dock, you could see hundreds of little white and gold crosses opening and closing as they moved along the surface.

When you think that these handsome and rather prominent creatures are only the reproductive phase of a thoroughly insignificant-looking parent hydroid, scarcely an inch long, drab and white, then you begin to wonder if there really is any order in the universe. We knew from watching them year after year that they would be around until March, and then, like so many other creatures, disappear.

Every winter large eolid nudibranchs appear. I watched one sweeping by under the light, fantastically wonderful with its white and blue striped hairs protruding from its orange body. It floated along upside down, clinging to the surface film, looking like a shaggy little blob. The two antennae that protruded from its anterior end reminded me of the large waxed mustache of a villain. I couldn't resist looking at it, and as it glided by I held out my hand and it was swept into my fingers. But when I lifted it out to examine it, it wilted into a formless lump of protoplasm and the little colorful papillae began to break off in my hand. I dropped it back into the water and the current swept it under the pilings and out of sight.

A penaeid shrimp, probably a stray from offshore, appeared for a moment with its eyes glowing like hot coals and then sank down into the depths. The wind gusted down viciously on me, and my wet fingers stung from the cold, and I was about to call it a night when suddenly I saw a batfish swimming by, heading right for the light and I hurried over to the the boat and grabbed a dip net.

At this time of year, any fish, especially a batfish, is at a premium. Usually we caught dozens of them in the shrimp nets, but they had been rare since the water temperature dropped. We could always find a customer who wanted a batfish for his aquarium, because it is one of the most peculiar-looking creatures in the sea.

In addition to its tapered body, its leathery skin, and its seal-like pectoral fins and froglike ventral fins, the batfish has a lure that it can project from its forehead. It has the same kind of lure that the primitive goosefish has, the anglers that sit upon the bottom and wave their lures at some unsuspecting fish or shrimp that comes forward eagerly to gobble it down, thinking it's a worm. In these angling fish the lure serves as a lure, and when the victim gets within range, the fish thrusts forward and snatches it up with its big toothy jaws.

But the batfish doesn't have toothy jaws. There isn't a tooth in its mouth, in fact it only has a very small round opening for a mouth. Even though I have sat on the dock and watched one trying to get hold of a small crab, awkwardly wobbling its body forward paddling its flippers from side to side, I have never actually seen it attack one.

When you cut open a batfish you don't find bits of shrimp antennae, you don't find the partially digested body of a minnow or crab, as you do in other fish. What you do find is a whole gut crammed with tiny snails and that's all. Snails, most of them less than a quarter of an inch long, tiny mud-dwelling mollusks so fine that the only way to find them in nature is to sift the sediments with a screen.

Since their guts never contained a heavy mixture of sand and mud, I assumed they didn't catch their prey by taking gulps of the muddy bottom. Detritus feeders such as croakers, mullet, and menhadens often ingest tiny snails, crustaceans, and worms while gobbling up the mud, but not the batfish. A graduate student at Florida State University who studied the batfish for his master's thesis theorized that it somehow used its lure like the Pied Piper to draw out the snails from the bottom. Perhaps the batfish swam into a snail-rich bottom, vibrated its lure or perhaps secreted some hormone or something, and when the snails popped out of the mud, the batfish selectively gobbled them up.

I reached out and caught it in my dip net, and decided to call it a night. It was almost two in the morning, and I was getting stiff and cold sitting there with the wintry night air settling down upon me. I really didn't want to drive up to the lab and put the fish away, but duty called. The wind and waves were making the floating dock bounce up and down, and I knew that tomorrow duty would call again when we had to go out and fill our order for a hundred *Nereis* for the National Cancer Institute.

The next morning, after reviewing the stack of orders that had to be shipped out that day, I told Doug and Edward that we were planning to go out in the bay and scrape the channel marker buoys to get *Nereis limbata*. Leon had gone off to Port St. Joe to buy some horseshoe crabs from the shrimp boats, since *Peneaus* was incapacitated, and the task of gathering the worms was left to us.

It wasn't a pleasant prospect. The bay looked cold and wintry as the icy winds bore down. The desolation of winter was upon us—the skies were cloudy and gray and the seas looked even more hostile and uninviting. The marshes around the bay were brown and shaggy, the only exceptions being some drab green clumps of sea ox-eye and the stubbly saltwort. The oaks that bordered the bay had lost all their leaves. Only the pine trees had any green, and on that day they too looked gray.

Beneath our slicker coats we had on sweaters and jackets, and our long underwear, but we knew that it wouldn't keep us warm. Doug wore his diving wetsuit, and he found that comfortable. "Come on," he said, loading the buckets into the boat. "Let's go out and get this mess over with. We've got a lot of orders to pack out when we get back."

He snatched the starter cord on the outboard motor but it refused to crank, and he pulled the cord again and again. Soon Doug and Edward were taking turns yanking the cord, and still it wouldn't start. "It's just too damn cold," he observed. When they were exhausted I tried it.

When I gave up, Doug started again, and finally the outboard spluttered, choked, quit, and then started. As we backed out of the boat stall, I looked reproachfully at the dock. It had let me down— it hadn't produced the worms we needed. I guess I had just grown to take it for granted.

Even still, docks are more agreeable than boats. They don't require an enormous amount of money to keep them up the way our shrimp boat did. Once we had raised its initial investment, it produced and produced for us. You didn't have to snatch a starter cord on a dock until it raised blisters on your hands. It didn't burn fuel, there was no lugging of gas tanks or worrying about faulty gas hoses. And there was no chance that the dock would break down while you were offshore and leave you drifting about in the freezing winds. The dock gave you the best animals in the sea, if you waited for them to swim past, and it gave you almost all the security of land. Praised be a dock, and cursed be a little boat in January.

We bounced along in the freezing sea, and when the tunnel boat plowed into a wave and sent the icy water showering over us, our teeth began to chatter and the wind pierced through our clothing. As we headed out we could see the bright orange sky in the east. The saying "red sky in the morning . . . sailors take warning . . . " has always held true. There were wind clouds high up in the atmosphere, and as we splashed and bounced and roared farther offshore, we could feel the wind getting stronger. It swept the sea in great gusts and made the waves rise higher and higher. I could see the Doug's fingers were red from the cold as he gripped the motor handle, his lips pursed. Edward and I sat miserably in the stern, watching the sea rise and fall, listening to the motor roar.

The farther out we went, the rougher and more unpleasant it became. When one big wave came crashing over the bow, we were suddenly standing in water up to our boots. We swooped up bucket after bucket of water and dumped it overboard. I called to Doug, "Let's call it off. The weather's getting worse by the minute. The damn worms aren't worth drowning over."

Doug pointed ahead. "We're almost there," he said, "I can see the cans."

It was true—we had just passed the last of the concrete channel markers that led into the bay. We didn't bother to try to scrape them because we had long ago learned that they didn't have the amphipod mulch that grew on the floating buoys that were anchored to the sea bottom by large concrete blocks and heavy chains. The fauna and flora of the concrete pilings were very similar to the life on our wooden pilings. They had their own zonation, beginning with a thin band of blackish green algae and then the same tiny fragile black barnacles, oysters, and sponges.

The bobbing cans were very much like out floating docks; anchored to the bottom, they rose and fell with the tide. But for some reason the fauna there was entirely different from what grew on our floating dock. These man-made floating steel cans, some

with lights, others with just numbers on them, marked the entrance to the Panacea Channel, and they were crawling with life. The steel surface, painted black or red, was solidly coated with tube-building amphipods, mostly *Jassa falcata* during the winter and *Coro-*

*phium louisianum* during the summer. Over the years we had come to rely heavily on these buoys to provide us with specimens.

Right after I bought *Penaeus* I found out what tremendous habitats they were. We were traveling far offshore, and came upon bell buoy 26, which clanged mournfully and told the seagoing barges that they were approaching land. I was expecting it to be covered with barnacles and I wanted to see what species might be living so far offshore. But when we eased up to it and Leon tied the bow of *Penaeus* to the big can, I thought it was one solid mass of brown stubbly algae. We raked a dip net over the surface and it came off in a huge solid sheet, and when I spread it out I found it to be one solid mass of amphipods. There were billions of them, literally billions.

*Jassa falcata* is a colonial amphipod that builds its tube out of mud and algae, and that habitat provides a home for a variety of other creatures. When I brought back a netful to our laboratory and let it set in a bucket for a while, hordes of flatworms began to crawl and ooze up the sides of the bucket. There were hundreds of long pink bristleworms, ribbon worms, small crabs, shrimp, and dozens of other creatures that we needed to fill our orders.

Once again, here was Gulf Specimen Company benefiting from man-made structures. We approached marker 8, a large black can without a light on it. It was a tricky maneuver with the waves rising and falling. Doug eased the boat up to it and Edward tried to get a rope through one of the mooring ears. The waves lifted our boat up and sent it crashing into the side of the can. Steel crunched into wood, but we held on for dear life. Doug and Edward scrabbled forward and grabbed the can with all their might and drew the boat up tight. Then they tied it as tightly as they could and held on while I scraped the can with my dip net. I scraped with all my might, pressing the rim of the net against the hard black steel and dragging it from the bottom of the buoy up toward the top. Suddenly the blue water was clouded with brown, as bits of amphipods, entrapped mud, algae, and barnacles washed through the dip net. The stuff peeled off, and we had half a netful, which we dumped out on deck.

Then I scraped more and got another half netful, and I could see where my scrapings had left a scar on the buoy as the life was stripped off and the bare metal lay exposed.

Again and again I yanked up the dip net, each time dumping it on the bow of our boat. If I kept it in the water too long, everything would wash through the mesh. Even though I hurried, when the net scraped into the living colony of creatures, I saw them explode and go into solution. I saw the flashes of pink as the worms we sought so desperately were lost to us and were probably gratefully received by the trout and pinfish below.

*Bang, bang, bang!* That buoy was murdering us. The boat and the buoy rose up on a swell and fell together, and it cracked into us. Holding on to that damn thing was a nightmare, the sea and sky all blended into one. The wind gusted harder than ever, and as we scraped, we formed an obstacle to the waves. The chilly water splashed over us, filling our boots, soaking us to the skin.

Finally we had enough mulch to examine, and I hoped we would be able to pick out a hundred worms and leave. We untied the boat and started drifting. In minutes we were being swept out to sea.

Seeing the cans disappearing behind us, I asked Doug, who looked slightly ill, "Don't you think we should drop anchor?"

He shook his head. "Let her drift. If we hold her down we'll rock twice as bad and I won't last five minutes."

The sea rose up and the sea fell down, and I had to admit that I was feeling a little queasy myself. I pushed that thought out of my mind, and went to work. We spread the mulch out and started going through it, and there weren't as many worms as I had hoped for. Sometimes you could pick out a hundred in a single net scoop, but this wasn't one of those times. They were easy to spot, their long sinuous pink bodies stood out sharply from the drab brown algae and mud mats, and while we picked through it our fingers were coated with the thousands of amphipods. The little gray specks were crustaceans, and they felt like fleas and they grew in incredible numbers. I felt a twinge of guilt about killing so many of these creatures just to get some worms. And there was no question about it, every one of them was doomed. We couldn't paste them back up on the buoy and we knew that if we dropped them overboard there would be a feast below, a feeding frenzy of small fish. So we decided to bring the mulch back and put it in our aquariums to provide our captive specimens with live food.

Earlier that week Anne and I had had a long discussion about taking life. Gulf Specimen Company takes in a substantial quantity of marine animals, both fish and invertebrates. In fact, everything that is taken out of the sea and put into plastic bags is doomed sooner or later. Death may be prolonged for months and even years in some cases by careful aquarists who take care of their specimens by carefully acclimating them to their tanks and feeding them. But sooner or later they will die, and even if they live, they have been removed from the food chain and from nature. When they are dead they are put in formaldehyde, thrown in the garbage, or flushed down the toilet. Nothing eats them, and their spawn will never go back out to sea.

Many didn't have a very prolonged life at all. I knew what was going to happen to the horseshoe crabs and fiddler crabs that we shipped. They would be lying on some laboratory table with electrodes embedded in their eyes, or used for some other research. The death there is prompter and perhaps more honest. Various human groups concerned with the welfare and suffering of animals have no objections to invertebrates and even fish being used as experimental animals. Their objection is to using higher forms, particularly cats and dogs, which have more-developed nervous systems.

The dividing line is subjective at best. I am certainly not in a position to say that the life of a single amphipod is less valuable than that of a spiny boxfish, or for that matter a whale. The closer an animal is to man, who believes himself to be the measure of all things in the universe, the more precious its life becomes. A handful containing a thousand dead amphipods didn't really upset me very much, but a thousand whales with harpoons sticking in their backs was another matter entirely. I could think of nothing more terrible

than the Japanese driving hundreds of porpoises to shore, banging
on pipes to jangle their sonar, and herding them like cattle to the
slaughter. I had seen movies of screeching, whimpering dolphins
being lifted out of the water by their tails fifty at a time bound
together by some monstrous machine while bloody butchers stabbed
their knives into their throats and ripped open their bodies. I was
sickened by it and gripped with rage at my fellow humans.

Yet as I watched the amphipods beginning to die in my bucket
I was objective about it, detached. They swam on the water frantically, and some were turning opaque and white, sinking to the
bottom. We had picked out thirty-seven polychaetes and still had
plenty of mulch to go. I skimmed the surface of the bucket and
looked at the hundred or so amphipods on my fingers. Life was
cheap. Nevertheless, it had occurred to me that maybe all life is
equal, be it amphipod, fish, porpoise, or man. If there is a life after
death, and if there is a creator of all this life, then maybe the quality
of an afterlife depends on how one has treated one's fellow creatures. Supposing there was a point against me every time I took a life,
just as you get points against your driver's license when you're
caught speeding. I would be in serious trouble. But then again, how
much trouble would the men who maintained the buoys be in?
Every year the Coast Guard cutters came in and lifted the buoys out
of the water with their heavy booms and replaced them with new
cans. The old encrusted cans were scraped down, sand-blasted and
painted with new black or red paint, and put back in place the
following year. The numbers of lives they obliterated, without
knowing or caring, were in the billions and trillions, and at that
point the numbers were meaningless. The new cans would sit in the
sea for a couple of months until all the antifoulant properties washed
out, and then life would start anew. In less than three months the
cans would be solidly massed with living creatures. Anne and I decided that it couldn't be much of a crime to do in specks of life if the
life is so rapidly replaced and in such teeming abundance.

I decided I had had enough philosophy when I felt myself
getting good and seasick. My face was flushed, my stomach was

turning sour, and I was wishing I was back on shore. I don't usually get seasick on small boats, but when I had to look down at those small creatures and concentrate on them, with the water rising and falling, I felt off balance and awful. I looked up and saw Doug gripping the rail and vomiting over the side. We had sixty-eight worms.

The buoys had to be scraped again before we met our quota of one hundred, but it was a nasty business. I hoped that the next time we had to come out here we would have calm seas and warmer weather. This miserable series of cold fronts, one right behind the other, couldn't last forever, I hoped as we headed back to the dock.

# 13 Shipworms and Sick Worms

It was February, and the last of the major cold fronts had pushed through and the weather turned unseasonably warm. All the gloomy overcast skies and the gusting winds had abruptly vanished. I stood on my dock enjoying the warm sunshine, watching fiddler crabs starting to come out of their burrows as they felt the sun's rays warming the chilled ground. All during the harsh freezes of the past few months there had been no sign of them. Even when I had walked over the salt flats where stubbly growths of pickleweed, *Salicornia,* and *Batis* grew and tens of thousands of fiddler crabs made their home, there hadn't been the first sign of a burrow.

A first-time visitor to the coast might walk the beach and declare that not even one fiddler crab lived in Panacea. But when the weather warmed up, those same salt flats and marshes at the edge of the dock were riddled with crab burrows. I watched them busily digging out their holes, molding pellets of mud and shoving them out the entrance. Some of the males were already standing at the entrance of their burrows, waving their single enlarged claws, rhythmically and methodically, hoping to attract some female that might be wandering by. Now and then a female would venture from her burrow, and the male would start waving all the more vigorously. Once in a while the female would hesitate for a moment and then dash down the burrow of her suitor. I could have spent many hours watching fiddler crabs, but that morning I didn't have time.

In a few hours the tide would be high enough and we were going to pull *Penaeus* out on dry dock. Lately she had been leaking worse than ever, and we could no longer afford to put it off. I walked out to the end of the dock where she sat, her bilge pump running, pumping the oily water out of her hull, spreading its shimmering iridescence on the crystal-clear waters. I hated being a polluter. I felt wretched about putting even that tiny amount of oil into the sea. Even with the most careful controls we would always have low-level spillage and waste. I hoped it could be minimized when the boat was finally hauled out and repaired.

On this exceptionally clear sunny morning I could see everything down below, including the hull of *Penaeus*. Her bottom was beginning to look like the rest of the Styrofoam floats; if anything, she was worse. Growing out from the once red-copper-painted bottom were big bushy masses of magnificently beautiful pink hydroids. Perhaps because she traveled out to higher-salinity waters and the mud and debris were periodically washed away, they were bigger than the hydroids on the pilings. It is really hard to describe hydroids. They branch and flower and expand, and are so delicate and fragile with their little pink polyps that they belong in a fairy world. And the barnacles, I could see them all over the hull. There were millions of them, feeding on the rising tide, throwing their feathery legs in and out like jack-in-the-boxes. It was amazing, our very own *Penaeus* was turning into a fouling community all by herself. She had clumps of green sponges on her stern, and even flattened gray oysters clinging to her hull. No wonder we couldn't run very fast. Barnacles can drag a boat's speed down to half and make it use forty percent more fuel.

It had been almost a year since she was last hauled out and had all that life scraped off. Shrimping had been poor lately, so she had sat unused and motionless, tied to my dock, and more growth had set in than ever.

Dickerson Bay was really amazing about growing things. That bay could grow more fouling growths faster than just about any other bay that I knew of. And the muds were much, much richer in life than in any adjacent bay. You could drop a bucket dredge down off the dock, scoop up some of the living mud and sift it through screens, and the most amazing crawling residue of amphipods, worms, snails, little clams, sea cucumbers, and other creatures would be strained out.

Certainly there wasn't another bay anywhere in north Florida that grew such an enormous number of turd sponges. They literally carpeted the bottom, looking like cobblestones. They seemed to appear by spontaneous generation. A month before, there hadn't been a single sponge on the flats—they had all died after the heavy rains the previous fall—and now I could see hundreds of the hard-packed, spongy elliptical blobs with their felted holdfasts. They were quite small now, scarcely an inch long, but by the time summer rolled around they would have grown to the size of a man's fist and would keep growing until the first winter freeze. After that, they would all die off and turn white and mushy.

It was a bit unnerving to stand on that very dock, looking down at that tide flat, which had been totally barren of sponges only a month ago. It was as if I were looking down at a huge army that had encamped overnight around my dock, as far as the eye could see. It looked as if they had *moved* in, but sponges aren't supposed to move.

The Panacea fishermen curse the sponges because when they "move in" they clog their nets and make net dragging impossible. But again, how could a sponge move? They have a highly modified felted holdfast that anchors them down into the mud, and I often wondered if they could somehow pull it up and get whisked along

*195*

with the current. I sent some off to Dr. Willard Hartman at Yale, and he identified them as *Tetilla laminaris*, but that really didn't help me a whole lot. I still didn't know why there were so many and where they came from.

Could it be that all the dissolved proteins that fed into Dicker-son Bay were nourishing the turd sponges? Scattered around Pana-cea were a number of crab-picking plants. When the thousands of blue crabs were steamed, shredded, and canned, the wash-down water from the crab houses gushed out of pipes into the bay. Often I watched schools of killifish hovering around the pipes in anticipa-tion of the bonanza of food, and hermit crabs would cluster all over the flats, chattering their mouth parts to partake of the nutrients that came by.

A little enrichment of a bay is probably a good thing. Too much and there's a chance of overenrichment and low oxygen conditions. Panacea's Dickerson Bay walked a fine line between the two. It wasn't polluted yet, but it could be, especially if land developers succeeded in filling in the wetlands. At the present, the marshes absorbed much of the nutrients.

As I looked at *Penaeus's* hull all overgrown with fouling or-ganisms that made their living by filtering their food from the water, I wondered if they weren't benefiting substantially from crab house wastes. There were great mats of pink anemones with greenish brown tentacles growing all over the floating docks. Normally a rather rare and obscure species, *Aiptasiomorpha texaensis* was only found sporadically around the coast, but here, off my floating dock, was probably the largest supply to be found anywhere.

I noticed that they too were growing on *Penaeus's* hull right on the waterline. What else was growing on *Penaeus*, I wondered, or should I wonder what was growing inside the lumber of *Penaeus*? I looked at the cross-members on the dock, the bracketings that were originally put down to lend strength to the pilings. They were coated with muddy black barnacles and oysters and eaten away by shipworms.

At a glance you could see that in many places most of the wood was missing, and the rest was so honeycombed with tubes that it would only be a matter of time before they disappeared. I knew very well that I would have to replace all the cross-ties one day and get a crew of men out there with fresh lumber and nails to hammer new stringers into place. My poor old dock needed a lot more work than that. We had done what we could for it after the tornado winds struck it last September, but it was still twisted and bent. It would take almost a thousand dollars to get it back into first-class condition, and we didn't have that kind of money for aesthetics.

The dock served its function, wounded or not. Boats could still tie up to her, and she still supported our oyster strings and fouling plaques for growing specimens. There were still years of life left in the pilings; they were still filled with creosote. Once they had been tall straight pine trees, then they were cut, trimmed, and soaked in a huge vat of noxious chemicals that completely saturated the wood, and then they were driven down into the ground to make a dock.

It was hard to say whether or not there were any isopods or shipworms in the pilings, because all those borers are amazingly resistant to chemicals. Even though the wood was soaked with noxious oils, marine growths still managed to settle on them and flourish. If you took a hammer and knocked off a chunk of oysters and exposed a black scar, shimmering oils would seep out of the scar at high tide. But in a few months the wound would once again be grown over.

You can't help but develop some admiration for these frail traveling larvae. Creosote is one of the harshest, nastiest substances I know of. When we were building the floating docks we had to put down a number of pilings ourselves, and when we carried them they inflicted long-lasting burns wherever they touched our skin. Watching the oily poisons leach out of the pilings and into the water made me feel even guiltier than pumping oil out of *Penaeus*. The pollution of the adjacent water was low-level but continuous. It varied with the seasons. In the summertime, when the water was warm, it oozed out of the pilings in little globs and droplets, washed away by the tides and currents into the vast sea, where it was diluted and mixed with all the other oil of the world. In the winter, when the water was cold, it became viscous and oozed out more slowly.

A scientist who was studying the fouling communities once told me that creosote was a carcinogen, and that it actually accumulated in the tissues of barnacles and other fouling organisms. When I asked him if it also accumulated in the tissues of the sheepshead, drum, and other large tasty fish that liked to hang around the pilings and munch on barnacles and small crabs and oysters, he just shrugged and said, "Could be."

Then I thought of all the delicious sheepshead that we caught off the dock and baked, and wondered if it was so good after all. I thought of all the other industrial docks, wharfs, and bridge pilings all over the world, all of them secreting creosote globules into the water for years and years to come, and I worried. Yet, if they weren't there, man could never build a wooden structure on the sea.

But even with all the creosote in the world, sooner or later the sea washes out enough of the surface oil and the pilings become coated with slime, and it is in this slime that higher forms of life can take root. If life is ever reborn or re-created on our polluted, overcrowded, creosoted, asphalt- and plastic-covered planet, we will have slime molds to thank. Some slimes that first cover the pilings are made up of bacteria and later filamentous algae. Sooner or later, no matter what object is placed in the sea in any part of the world, slime begins to cover it.

To a boat owner, that first bit of life is a liability, but to a specimen company it is a tremendous asset. We deliberately grew shipworms off our dock so that zoology classes could study them. We cut some four-inch squares of plywood, drilled holes in their centers, and then strung them on ropes and hung them off the dock. In six months or so the wood would become riddled with worms, and when we shipped them off to a zoology class in Wisconsin that had previously only read about shipworms, they could split the wood apart and actually see the worms eating the wood like termites. It is an unforgettable sight especially to someone who has only seen logs and boards slowly deteriorating in fresh-water ponds, an endless process that takes years and years.

We found that by treating the surfaces of these wooden plaques with various antifoulant substances we could alter the life that settled on them. For example, when we used the same copper paint that we used on *Penaeus*, the plaques stayed absolutely barren of life for almost a year, and then the slimes took over. Barnacles didn't grow on the painted surface, although they completely caked over the untreated plywood squares. However, since the barnacles were repelled, there was no competition for bryozoans, which resisted the paint, took over, and flourished. After a while, those same boards had a pure culture of white slipper limpets, *Crepidula convexa*, and nowhere else on the dock could these be found.

We could grow all sorts of animals on plastic squares, and we even grew clusters of brown *Padina* algae when we drilled holes in chunks of limerock and strung them off the dock. Never before had *Padina* been seen in Dickerson Bay. It was found offshore on limerock bottoms, but never inshore. But if we wanted wood borers, then we had only to hang out untreated wood, and we "cultured" all we needed.

If we didn't harvest the plaques and ship them off, then they would be eaten away to nothing and would have to be replaced with new ones. After a while we could mush up the soft board between our fingers. Shipworms have had their role in history. Every mighty sailing vessel that has crossed the sea and explored the new world, and every wooden bridge that ever sat in brackish or salt water, has crumbled and finally disappeared because of the tiny, busy wood-boring and wood-eating organisms.

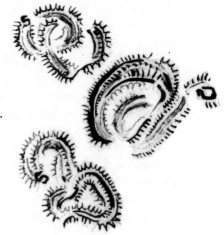

There is a difference between the two. Shipworms such as *Bankia gouldi* and *Teredo navalis*, which live so abundantly along the Atlantic and Gulf coasts, are borers and do not eat the wood. They use the lumber to make their homes, building their lime-coated tubes within it, expanding their feathery siphons into the sea to partake of the plankton.

They are not worms at all, but very primitive clams, and their bivalved shells, which are small, are highly modified with teeth that rasp away at the wood as they twist and turn until their burrows are made. The shell is only a tiny portion of the clam's great elongated body, which may stretch a foot or more through the wood, out to the surface where the clam can feed. Perhaps to make their burrows more comfortable, or to lend them support, they secrete a lime coating. So when a student breaks apart a board, he will see dozens of limy white tubes and large fleshy white worms inside them. Often the worms will die off, leaving behind sand-filled tubes of lime.

Then there are the true wood eaters, which are perhaps even nastier. *Limnoria tripunctata* are isopods, related to the pill bugs that you find in a garden under a rock, or the gray sea roaches that scuttle out in all directions when you walk out on a dock. When you see a large wharf piling that is eaten down to a point, like someone put it into a gigantic pencil sharpener, then you know you are looking at the work of *Limnoria*. Fortunately, none of our pilings looked like that, but the pilings of the older Rock Landing Dock, where the crabbers tied their boats, were all badly eaten and the dock wobbled dangerously when you walked out on it.

If I wanted some to ship, I usually walked out beneath the older dock at low tide, grabbed a hunk of the piling and tore it off with my hand. It was easy to do. When I looked carefully, I could see that the piece of wood was riddled with hundreds and sometimes thousands of tiny holes. The wood itself was usually muddy from the silt suspended in the water that lodged down in the burrows, and it was practically impossible to see the *Limnoria* with my naked eye. But when I took the wood back to our laboratory and put it under a dissecting scope, I could get a good view. Then I could actually see the tiny little bugs down in their holes, eating away.

According to Dr. Robert J. Menzies of Florida State University, *Limnoria* are the only sea creatures that can actually digest cellulose.

The little isopods have a highly developed pair of mandibles, the right one with a rasp and the left with a file. As they saw away at the wood, fragments are crushed up and swallowed. Menzies's studies have shown that *Limnoria* can eat their dry weight of wood in ten days, which puts termites to shame.

They are prolific little creatures, very adept at reproducing. Females carry their young in brood pouches until they are ready to make their way in the world; then they leave the pouch as miniature adults, swimming off at night in search of new wood. This form of reproduction has a tremendous advantage over that of creatures which must live in a planktonic state, undergoing numerous stages of metamorphosis before they can mature. Survival prospects for the little isopod are greatly enhanced. It can live for a full month swimming in the sea, looking for a home.

The requirements for *Limnoria* homesteading are not as easy as one would believe, however. They are gregarious creatures and prefer to live in wood that is already infested with others of their kind. In fact, if an isopod lands on a surface and eats its burrow, and after a while no other isopods come to stay, it will leave and go elsewhere to find company. Sometimes, too, they will leave a piece of wood that has been almost devoured, and move elsewhere.

Most of this migratory movement takes place at night. The males leave first, and when they are established in a new piece of wood, the females join them. They arrive swimming, and crawl over the surface, going from burrow to burrow to see if a male is there, poking their antennae down inside. If the burrow is occupied by the opposite sex they disappear into their new home to raise a family. They are amazingly immune to antifoulant paint; all they need is one little scraped area where the surface is exposed and they can move in and take over.

Needless to say, I had very mixed feelings about isopods and shipworms. Here my boat bottom might be riddled with them and it would take hundreds of dollars to fix the bottom. As I stepped aboard *Penaeus* and looked at the oily water rising from the floorboards over the engine cavity, I had less and less use for them.

For all I knew, when *Penaeus* was finally hauled out and the barnacles were scraped off, I'd be able to drive my foot through her stout lumber. I didn't think that would be the case, but one never knew. Some years there would be bumper crops of isopods—all it took was the right combination of salinity, temperature, and nutrients, and there they would be.

Yet I knew that they certainly had their place in the design of nature. They cleaned the rivers and estuaries of wood. Rivers help nourish the sea by dumping leaves, bark, stems, and branches of upland vegetation into the fast-moving currents that find their way to the sea. They also carry down huge logs and often entire trees that are washed away by the current slicing into the riverbank, and sooner or later, it may take years, it all ends up in the sea. Beaches and shorelines erode, exposing the roots of pine trees, which die and fall into the water. Hurricanes rip down the mighty, twisted, windblown forested shorelines. If it were not for the busy little bivalves riddling their burrows into the great oaks and pines and cypress logs, and all the other little creatures that attach to the wood and dig their homes in it, the ocean would soon be peppered with logs.

As I stood on the dock I looked up at the head of the bay where the old *Isabel* now sat high and dry on the mud flat. Even from that distance I could see that planks were missing from her oyster-grown bottom, and I knew that shipworms were doing a job on her hull. All the remaining metal on her deck was a rusty orange, and in a few years, between the worms and the salt, there wouldn't be any remnant of her hull above the tide flat. Someday there might be a big oyster bar there, or even a marshy projection of the shoreline, but nothing more.

Looking at *Penaeus*, I knew that someday she too would join the boneyard. We could prolong her life for a number of years, but that was all. For the moment anyway, *Penaeus* belonged not to the worms but to us. And that meant that the worms, the barnacles, the hydroids, and all the millions of living forms that had taken up residence on her hull had to go. I finished gathering the specimens I had come after and then returned to the lab. The tide had risen.

It was time to haul out *Penaeus*. It was eleven-thirty. She was scheduled for pulling at noon, when the water was tiptop high, far into the grasses. The weather couldn't have been better, with the sun shining and a gentle breeze blowing. Once the hull was scraped, *Penaeus* would be ready for painting. With luck, we could have her back in the water in a week.

We towed her down the bay with the tunnel boat and anchored her off the old Panacea Ways. The boat ways at Panacea had seen better days. Years ago, when Panacea was a bustling town and there were many things going on, and the fishing industry was much greater than it is today, the boat ways were a busy place, with a younger Baisden Roberts and other men hauling boats out and repairing their bottoms. But now the area that had been the boat ways was almost completely reclaimed by nature, and sand live oak, yaupon, and gallberry grew up around the railway. It was a peaceful place now, with towering slash pines overhead and woodpeckers hammering away, not a bustling boatyard.

It was against this setting that *Penaeus* was being hauled up on rusty railroad tracks, on a wooden cart with orange rusty wheels. The ancient winch clattered away, pulling in the loose rusty cables, and Baisden, the old man at the controls, looked intense and thoughtful as he operated the groaning old piece of machinery. Inch by inch *Penaeus* crawled up out of the bay and on to the elevated railway, and finally she inched into the marsh grass that grew around the railings, knocking the periwinkle snails off the tops of the marsh grass where they crawled up to escape the high tide.

The red copper paint on her waterline was now sitting high out of the water and as she came up farther, the shaggy barnacle-covered hull became exposed. You couldn't see a bit of the original copper paint, it looked like it was one solid sea of clicking, bubbling barnacles that were exposed to the dry air.

Hundreds of thousands of barnacles and tufts of hydroids were everywhere, hanging limply and forlornly, and clumps of oysters covered the hull. Just the numbers of animals that had attached and grown on the bottom was enough to make one stare with wonder.

*Click-click-click*, the rhythmic sound of the winch continued, and more and more of *Penaeus* was exposed to the air as she inched up the incline. Soon there was no water beneath her hull.

The bubbling and hissing of the barnacles was practically deafening under the boat; it was as if they knew that their whole world had come to an end. The year that they had lived on the boat, the miles and miles of water they had traveled, the plankton they had sucked up—all of it was to be a thing of the past. Even though the Panacea Boat Ways hadn't been used in ages, the ground beneath the working platform had a thick mat of ancient dead barnacles that had long ago been scraped off and packed into the sandy soils.

Three men dug into the barnacles with their heavy scrapers and the barnacle sounds seemed to increase. It was hard work, raking them off in sheets, scraping their metal blades back and forth, crushing the shells, tearing them off their bases. Leon and I hurried around behind them, catching the clumps of barnacles in buckets. I grabbed a handful of the mashed barnacles and looked them over. There was no doubt about it, there would be thousands of *Nereis limbata* on that boat bottom.

"Look at these," cried Leon excitedly, holding his hand out with two big pink worms for my inspection. "We'll get a hundred in no time. That man will go stark staring crazy when he sees them. He'll want a thousand like them. By God, *Penaeus* even makes money when she's up on the ways."

"How much you get for them worms?" Henry asked me as he stopped his scraping to look at them.

"He gets plenty," said Baisden a little grumpily, "you can bet on that."

As the scrapers continued cutting great swaths on the bottom, the red paint was becoming exposed, and soon there were whole areas that were cleared, with only white barnacle scars left behind. There is a rule that one abides by when pulling any boat. As soon as the boat is hauled out, you have to scrape it right away while things are still alive and soft. If they are allowed to dry out and desiccate, they become firmly fixed and turn into a messy glue.

At last we had our buckets filled with barnacle mush. Most of it was shells and tissues and yellow globs of eggs. Leon and I decided to sit beneath the shade tree and pick out the worms rather than haul them up to the lab. There was better light down there, so we could see them more clearly.

I went down to the marsh and fetched a bucket of water. The tide had dropped since they had begun scraping, and the mud flats were beginning to expose. As I waded out over the shallows I noticed a number of fiddler crabs down by the water's edge and I made a mental note to round up some before long. The warm weather was too good to last. A cold snap would set in before long, and the fiddlers would disappear again.

I trudged up the beach, past the men working on the boat, and poured some water into Leon's bucket of barnacle mulch. Immediately the water turned orange, coppery orange, and I had a sinking feeling.

"I reckon that was a dumb idea after all," said Leon, looking at the poisonous red water. "All that red paint that come off with the barnacles is going to kill off everything here."

We culled the hash, picking out the long pink worms. They were barely moving. Normally they squirmed and crawled around with vigor, but now they were as limp as spaghetti. They looked rather bleached out, and their proboscises were open and inverted. So there it was; beneath all the slime and the bases of barnacles was poisonous copper paint, still capable of killing.

Before long the entire hull of *Penaeus* had been scraped. The men hosed it down, and the water that ran off the hull was bloody red from the copper paint. The water washed away the residue, bits of shell and mushy tissues, exposing the neat round calcareous rings left by the thousands of barnacles. Those rings would be much harder to scrape off. The old men were sweating as they racked their scrapers back and forth and washed more red paint off. Finally, Baisden was finished. He drilled a hole into the hull and all the oily black water gushed out onto the ground.

I watched the copper paint running off into the marsh, mixing with the bay water. I hated that. Over the years, millions upon millions of gallons of this highly poisonous substance have been washed into the sea from boatyards all over the globe. I had bought many gallons in my lifetime and had painted it on many boats. The stuff we planned to put on Penaeus this time was one of the more expensive brands, almost guaranteed to keep fouling growths from coming back for a much longer period than the cheaper brands. But again, what could I do about it? Haul that monstrous boat up on a trailer and take it five miles inland and scrape it down on sand-hill country? That was too impractical and too expensive.

I watched the copper paint mixing with the water, spreading out. And then I saw the fiddler crabs near the water, basking in the sunshine. But at a glance I could see that something was wrong. They didn't move like fiddlers do, they were wandering around as if they were a bit goofy and drunk. Before, when I had walked past them, they scuttled out of the way and some stood cautiously at the mouths of their burrows. Now I could see that they were sick. Some were falling over on their backs and twitching their legs spasmodically.

"Can't we dike this stuff or do something?" I asked Baisden as he continued washing. "It's killing all the fiddler crabs."

"So what," he demanded defiantly, "there's plenty more, and they ain't good for *nothing* no-how. I swear, nowadays you can't do *nothing* without messing up the ee-cology. What do you want people to do, quit living? We used to kill them by the millions a few years ago when the boatyard was working, but shoot, there's plenty left to take their place. That's the trouble nowadays, all these people bitching about this ecology stuff is ruining the country. You better go to worrying about all these worms you got in this boat."

He pointed to a number of tiny pinholes that were peering out at me from the boards. Leon inspected them and shook his head. "No damn wonder she's leaking. Some of those boards are gonna have to come out."

"Now I can plug up some of them, you know, and drive a cypress plug in them, but some of these others—that's a different matter.

Now that's where some of the water is coming in from, but the rest is coming in from all this rotten caulking in the seams. All that will have to come out and be replaced, too."

I watched the poisoned fiddler crabs walking around in a dizzying circle "Well, how much is that going to cost?"

"Well, I don't rightly know," he said, rubbing his chin. "Like I told you, we do it by the hour. Why, if I was to tell you one thing and it came out another, you'd get mad. So I'll just tell you that we work for four dollars apiece, so that's twelve dollars an hour. A boatyard would cost you a heap more."

"Well, how many hours do you estimate?" I demanded. "Five? Ten? How many?" I was getting anxious. I hadn't bargained on the extra expense.

"It shouldn't be over twenty hours at the most," he said. "Not if all of us get right on it."

"All right all right, but when will she be ready?"

"I'd let her stay up there for a week or so and dry out good. We'll tear out the boards tomorrow and then we can get in and work her over. Now I told Leon here that we'd be able to pull and clean her today so you-all could paint her. I didn't figure on doing no big amount of work. We're building a room onto old lady Johnson's house and all, and we got to finish that up, but we'll get on this just as soon as we can. You know how it is. . . . "

# Death of a Porpoise

Nowhere can you better observe the changing seasons of the sea than standing on a dock, looking into the water day in and day out, month after month, year after year. This dock, with its wooden plankings and Styrofoam floats and creosoted pilings stands rooted firmly in the ground beneath the water, unchanging as the world changes around it. It merges the world of the water with the world of the land.

Different life comes in with the changing seasons, and the oceans, bays, and estuaries all have their seasons, just as the woodlands and forests have theirs. While the marigolds are blooming on land, pink *Eudendrium* hydroids are blooming on the pilings and schools of threadfin herring and menhaden come in from the deep and glitter in the water. As the robins suddenly appear fluttering and chirping in the treetops on shore, the waters in the Gulf are suddenly filled with moon jellyfish, bobbing and glimmering in the sunlight.

As the sea water temperature began to slowly rise, porpoises, *Tursiops truncatus*, began appearing off the dock once again, joyfully charging into the midst of schooling mullet, leaping high out of the water and landing with a mighty splash. The cold had driven them offshore, along with their food, and now that the bay waters were beginning to warm, they were back, swimming in circles, rubbing up against each other and thinking of love.

Just watching them brightened my spirits. Even though the water was still cold—I felt that the porpoises were harbingers of spring and warm weather.

Spring in the sea begins slowly, subtly. Jellyfish were pulsating into the bay, and there were a few minnows. As the days grew longer, the sun shone down upon the mud flats and the music of life began to increase slowly in tempo. It was like the melody of Ravel's *Bolero*, beginning softly and slowly and then gradually picking up and going on and on with vigor and purpose. The marsh grass began to look greener. The clapper rails were beginning their courtship, and horseshoe crabs were stirring from their winter homes down in the mud and starting toward shore on their mating migrations.

Soon the water would be getting murky—it would take only a few weeks of warm sunshine to make the tiny plant cells start to divide and grow, and then billions of little copepods and arrowworms would appear, and before long the bay's nutrient-rich waters would be a living soup. But for the moment it was still clear, just as crystal clear as it could be. Standing on the dock, I could see the coppery red painted hull of *Penaeus*. Somehow she looked strangely naked without her bearded growths, but I was glad they were gone and hoped they would never come back.. By the time eight worm-infested boards had been ripped from her hull and replaced with new ones, I had a bill of close to seven hundred dollars.

I preferred not to think about it. Soon it would be spring and the shrimp would be running, and maybe that would help us pay for the repairs. Anyway, it was too magnificent a morning to be involved in mundane things like paying bills. Down beneath the glassy waters I could see life scurrying about. Menhaden minnows were all over the dock, feeding among the hydroids, catching the sunlight and glittering.

I watched a large white flat-clawed hermit crab dragging its whelk shell across the mud, making a serrated trail, crawling over the turd sponges that covered the flat. What a great feeling it was to sit there and look down into the sea and see everything.

I often wished it could be that way in the summer as well as the winter, instead of the turbid soupy mess that it was. Then I would be able to look down at the encrusted pilings and see trout feeding among the fouling organisms. I would be able to see schools of spadefish and croakers foraging, and watch them darting in and out, grabbing up mouthfuls of mud, spitting them out and keeping the edibles.

Beneath the dock, next to the rocks were a number of stone-crab burrows. I could see that as plain as if I were standing on top of them. One burrow had a big crab sitting at the entrance, with its two massive claws protruding slightly. I watched the crab for a few minutes; it wouldn't come out of its burrow. The commercial stone-crab fishermen often told me that their catches declined drastically when the water cleared up. Stone crabs don't like clear water, they love murky turbid water, which hides them from their predators and allows them to get out of their homes and move about freely, crushing oysters in one big claw as if they were peanut shells, and delicately picking out the meat with the other. It's hard to say what kind of predator other than man would eat the stone crab, with its hard shell and huge grinding claws. Perhaps the loggerhead sea turtle could crush its hard shell with its powerful grinding jaws. Loggerheads often frequented our bay, and I knew that soon they would be coming inshore and lumbering out of the surf at night to lay their eggs. Although they dearly love blue crabs, and often drive the crabbers to fits of anger by grabbing their crab traps and twisting them into useless little balls of wire, they also love to feast upon jellyfish, particularly the stinging nettles *Chrysaora quinquecirrha*, which had drifted into the bay in shoals.

The clear waters around the dock were full of the nettles, pink parachutes that pulsated slowly along with a flowing network of stinging tentacles. Their soft pinkish brown bodies had striped markings that resembled the spokes of a wheel, and they were magnificent to watch. As the tide was rising and swelling into the bay, I saw a dozen drift past in the current, then fifty, then a hundred. In fact, they were all around the docks, pulsating, contracting, expanding. It was a magnificent sight.

I knew that sooner or later some biochemist would call and ask if we could supply him with frozen jellyfish so that he could study their protein stinging cells, and I headed for the lab to get my crew so we could come back and stock up.

I drove into the yard and met Leon coming out of the office. "I was just going to get you," he said excitedly. "Some woman on Alligator Point just called, not two minutes ago. She said there was a wounded porpoise on the beach in front of her house."

As Leon, Doug, and I sped down the highway to Alligator Point, I couldn't help thinking the worst. I was sure that some angry crabber had blasted a hole through one of the porpoises that was following his boat. Lately my friends and neighbors had been complaining more and more about porpoises, insisting that they were stealing the bait and turning over the traps. No, they weren't sharks and they weren't sea turtles, they were porpoises, they would say. During the winter the porpoises had a hard time finding fish, and those freshly thawed mullet, stuffed into traps and dumped overboard, were too much of a temptation.

They saw the porpoises bobbing and playing around the boat, and to be double sure, they would throw the baited trap out, watch a porpoise dive down and snatch the cork, and then when they retrieved the trap a moment later, it would be empty of bait. Unbaited traps don't catch crabs, so a man lost both ways.

I often saw frustrated, angry men come in at the end of a day's fishing with three or four hundred pounds less than what they should have caught. They appealed to me to help, and I wrote the Department of Natural Resources a long memo and followed up with several phone calls suggesting that they start investigating the problem. Maybe some kind of sonar buoy could be developed, or maybe they could help the fishermen design a crab trap or a bait cage that would be porpoise-proof.

The Department of Natural Resources responded by calling to tell me that all those fishermen were mistaken. The traps were obviously being bothered by turtles or sharks, but not porpoises. A porpoise had too sensitive a nose to open a wire trap.

As I drove down the highway, I talked it over with Leon. "Now ain't that some stupidness," he scoffed. "That goes to show you what the damn state knows. If they ever had crab traps, they'd know what a shark does to them. They tear them all to pieces, and a turtle, he bites the wire and then tears it with his flipper. But I seen porpoises do it, they don't do nothing but flip the trap open, pull the bait lid open and snatch out the fish. They're smart damn things."

"Well, do you suppose someone shot this porpoise?"

"It could be. There's a lot of mad crabbers out there," he said, as we sped past the endless clutter of beach houses that stood row after row, blocking the sea from our view.

All we could see were little patches of blue water between the structures. The woman had given us directions, but all the houses looked alike.

The beach area was almost barren of people. In the summertime it would be one endless sea of humanity and bumper-to-bumper traffic, but the weather was still a little chilly, and only a few people were swimming. So when we saw a small crowd of people standing around a small stretch of open beach, looking into the water, we knew we were in the right place.

I hurried behind Leon and Doug and there before us was one of the biggest porpoises I had ever seen. It was lying limply in the waves, with its curved dorsal fin protruding above the swells. All we could see of it was its long dark shape. There were no wounds visible. There wasn't the first sign of movement. It was as limp as death.

A woman came up to us. "Are you with the specimen company?"

"Yes," I replied, "Are you sure that porpoise was alive when he came in?"

"He certainly was," she said. Several other people nodded in agreement. "He was thrashing and jumping around on the shore and blood was coming out all over the place. I've never seen so much blood. It looked like something bit it."

From where we were standing, I couldn't see any blood. It had all washed away in the forty-five minutes that had elapsed since she had called us. "Let's get its head out of the water," said Leon. "That porpoise is bad hurt. He'll drown lying there if he ain't done it already."

I looked over its long rubbery-looking body lolling to and fro with the waves. We waded out to it, grabbed it by its two pectoral fins, and lifted it up. In three feet of water the heavy animal was fairly buoyant. We strained hard and lifted its head out. "Oh Jesus," cried Doug, "look at that!"

Before us was one of the most gruesome sights I have ever seen, a terrible gaping wound in the throat. A large chunk of flesh had been ripped out and it was still trickling blood. The porpoise emitted a sickening wheezing gasp through its blowhole. We dragged it with all our might up onto the beach, as far as it would go until it was hard aground and its bulky weight wouldn't let us drag it any farther. It must have weighed close to seven or eight hundred pounds.

The great mammal let out one final gasp, blowing a stream of hot vapory spray out of its blowhole again, quivered, and went limp. The last terrible moments of the porpoise's life were finally over, and now it was just so much dead meat. The gruesome wounds were all over its body, where razor-sharp teeth had gnashed and torn and scraped it. There was no question about what had killed it—a large shark.

The crowd moved closer; there were looks of horror and revulsion on their faces and strongest of all was fear. We all felt fear and awe as we looked at the chunk of flesh that had been ripped out of the rubbery skin in one powerful bite. When we rolled the porpoise over we saw more bloody red serrated teeth marks torn through the bluish black hide. Clamminess ran through me. I involuntarily stepped out of the water, even though I had only been standing in water up to my knees.

I found myself gazing out over the horizon. Somewhere out there, perhaps only a few hundred feet away, was a shark large enough to have struck down this mighty nine-foot porpoise. There weren't supposed to be sharks inshore at this time of year. It was too cold for the common species. What kind of shark was out there? A big tiger shark? A hammerhead, thirteen or fourteen feet long, or was it some big pelagic species that had wandered inshore, something like a mako or a great white?

The bite was less than a foot across, but considering the smoothness of the porpoise's skin, the firm rubbery curvature, a moving porpoise didn't present a surface to get in a really good bite, certainly not like an arm or leg or human torso. But the bite was effective, and aimed right for the throat on the first shot. Not more than two or three hours ago a mighty set of jaws had crunched down and a heavy powerful body behind them wrenched to and fro and tore away the flesh. Then the wounded frightened animal swam frantically, gushing blood and heading for the shallows to escape his tormentor. Or was it tormentors? There were bites on its stomach, and tears and teeth marks on its back.

What was the story behind this terrible incident? The porpoise certainly didn't look sick. There were no old festering wounds or abrasions. I checked it over very carefully, went over it inch by inch to see if there were any sign of a bullet wound, but there was none. Aside from the bites, this porpoise appeared to be the picture of health. It had clear eyes with no sign of infection. I checked its long snout and white pointed teeth. This was no helpless calf, this was as big a porpoise as you could find anywhere.

A young man crouched down next to me and looked it over, touching the raw cavity. "I thought porpoises were supposed to kill sharks. I read someplace that they butt the sharks with their noses and batter them to death. It's supposed to be the other way around."

I shook my head. I didn't know. I had read popular accounts of a band of porpoises ringing a hammerhead and in a cooperative effort smashing into it with their noses until they caused enough internal injuries to send the big monster spiraling to the bottom. I once met a shrimp-boat captain who had witnessed something similar, but here before us was proof to the contrary.

Several years ago during the summer, a wounded porpoise swam into Dickerson Bay and beached itself on the shallow mud flat directly across from our dock. Following at a good distance behind were two six-foot sand sharks, but they made no attempt to attack. It appeared that they were only following like vultures, waiting for the porpoise to die. Leon and I had gone out to investigate. There were infected sores all over its body. Its eyes were filled with mucus and running, and its lungs were badly congested. There was a bullet hole in its back. By the time I was able to reach any state authorities, the tide had come in and the sick porpoise had swum off with his following of undertakers.

Until I looked at the dead adult porpoise on the beach, I assumed that only sick, weak, or dying porpoises were attacked by sharks. In their capacity as sanitation engineers, sharks culled the ocean of the diseased and the weak and the suffering. They were predators, just like wolves and mountain lions on land, which trimmed the soft and weak deer.

Perhaps this porpoise was different. Or maybe it had a brain tumor or some disease that I couldn't recognize. It certainly looked healthy enough, but only a veterinarian could really tell. Who knew what would be found when the carcass was autopsied? Maybe it was a pregnant female, but I couldn't even tell the sex. However, I could bring it to specialists who could.

My mind began to reel. This was truly a valuable scientific find, because here, for the first time, was documented scientific evidence of interaction between sharks and porpoises in the wild. For years

Dr. Perry Gilbert of the Mote Marine Laboratory in Sarasota, Florida, had been studying their interactions in captivity and found the results varied and often inconclusive. Sometimes the porpoises would advance on the sharks, making clicking aggressive noises; at other times porpoises were observed to be intimidated by sharks. But the more normal reaction was that sharks and porpoises ignored each other. Dr. Gilbert stated on many occasions that there was a tremendous amount to be learned before any statements could be made.

The mystery of what happened to the porpoise that died on the beach of Alligator Point was never solved. It was never given a chance. No veterinarian was ever allowed to autopsy the porpoise to see if it was somehow diseased. No shark specialist ever probed and measured the wounds, nor measured the angle of the teeth marks to see what species had attacked it. Our important find was lost to science forever, and it was all my fault *for obeying the law.*

We loaded the porpoise into the truck, helped by the crowd of spectators, and sped back to the laboratory. I was planning to call Dr. Gilbert and make arrangements to deliver the porpoise to his laboratory. Filled with excitement and enthusiasm, I first called the Department of Natural Resources, to comply with the law. They were in charge of issuing scientific permits and our permits prohibited us from taking marine mammals. After going through the switchboards of various departments, I finally got the officer in charge of marine mammals and excitedly told him about our find and our plans to haul the porpoise down to the Mote Marine Laboratory for study. I also offered to provide all the data to the state's marine laboratory.

But the voice on the other end was cold and official. The commanding officer informed me that I was in strict violation of state and federal laws and that for possessing a marine mammal I could be fined five thousand dollars and receive a prison sentence.

"But wait a minute . . . this is for scientific purposes," I protested. "We're not going to eat the damn thing. This is the first documented case of a —"

"I'm sorry. Unless you have a permit from the U.S. government and have advertised in the Federal Register, you are in violation of the law. You said that you have the porpoise in your possession?"

"Yes . . . " I replied uneasily. "It's at our laboratory right now, in the back of our truck."

"Well, seeing that you have called us, it's obvious that you are acting in good faith. I don't think we'll prosecute or that any action will be taken against you if you continue to cooperate."

"I'm more than willing to cooperate. Why don't you have your marine laboratory take it then?"

"No," said the voice of authority. "The law is very strict. And we are required to enforce it for the National Marine Fisheries Service of the Department of Commerce. The animal must be taken to the nearest sanitary landfill and disposed of by the public health department. Just stand by. We'll send an officer down in a little while to see that it's properly buried."

I was in a hopeless rage. For a while I thought of hauling it up to the giant blast freezer at the crab house where they kept their fish bait and then calling my attorney, congressmen, the press—the whole works—and see if we couldn't fight it out. I called Dr. Gilbert. He asked that we try to cut out the wound tissues, preserve them in formaldehyde, and send them to him. But he could offer no help against the bureaucracy. He was having his own problems with it. I didn't have the time and money to fight the government, and a battle like that could do only more harm than good toward protecting marine mammals.

Anne came out of the office to look at the porpoise while we waited for the marine patrol to arrive. "Isn't it ironic that just a couple of years ago we were writing letters urging Congress to pass the Marine Mammals Protection Act? Who would have thought it would come to this?"

"It's still a good law, Anne," I said miserably. "Even if it does stop people from utilizing dead porpoises for scientific purposes.. .."

"A great law," she agreed. "The American tuna fishermen are drowning dolphins by the thousands out in the Pacific, and they're exempt."

"That's true," I admitted, "but it does stop whale products from coming into the United States."

"It was a good law when it started, but it's been subverted by bureaucracy," she said, climbing up on the tailgate and looking down at the porpoise. Then she saw the gruesome bite on the underside near the throat. "God, is that ever an ugly wound. Some big hungry really got a good mouthful there."

"How do you feel about diving out there now? It probably happened where you and the rest of the FSU oceanography people were diving a few weeks ago." Anne frequently went on offshore survey trips, sometimes as far as fifty miles out in the Gulf of Mexico, diving down eighty to a hundred fifty feet to collect samples.

"You can't worry about things like that," she said, shaking her head. "If you think about them then you'll be afraid to dive, and diving is too wonderful an experience to miss. If it happens, it happens, but the chances are it never will. Hadn't you better get the camera and take some pictures of the wounds, and get the best measurements we can? Maybe Dr. Gilbert can make something out of them."

Leon helped us turn the porpoise and measure the wounds and bites while I blasted off two rolls of film. Leon was grumbling, "It's the goddamn craziest thing I've ever heard of in my life. To think we went to this much trouble to haul that thing up here. It must weigh close to a thousand pounds. We damn near broke the axle on the truck driving over the beach with that much weight in back and now we got to haul it to the dump. . . ."

The sleek gray and black marine-patrol car with its blue dome light pulled up into the yard. Leon was filled with the same frustration that I was, and he snarled, "Here comes that sorry old son of a bitch Wilson. I wonder how they managed to get him out of the coffee shop?"

Officer Wilson was courteous and respectful. He ignored Leon and inspected the porpoise. "Looks to me like a boat propeller chopped it up. Sure is a shame."

Leon spluttered his anger. "You goddamn peanut farmer . . . if you can't tell a shark bite when you see one, you ought to go back to your mule and plow instead of being on the state's gravy train."

I didn't want to antagonize the officer. I had hopes that by being pleasant and cooperative we could still get samples for Dr. Gilbert. "Look, Wilson, I think they're bites too. If we're going to bury it anyway, why not let me cut some skin off around the wounds and put it in formaldehyde. Then you can bring it to your research department and they can ship it to Dr. Gilbert, and we can find out for sure."

"We don't want to be uncooperative," he said, "and you know yourself that we have a good research laboratory. But we're acting as federal officials and obeying the law ourselves. I've got my orders—no part, pieces, tooth, or organs may be retained by any individual unless they hold a federal permit, and that's the law."

"Bullshit!" snapped Leon, who was formerly a commercial fisherman and therefore despised the marine patrol. "You're just being plain old shitty."

I told Leon to calm down and go pack the orders. He stomped off grumbling. I turned to the officer and apologized for Leon, still trying to get samples of the wounds, but he wouldn't permit it.

And so, with Doug driving, followed by the marine patrol, we headed off for the city dump. When we arrived at the stinking "sanitary" landfill, with its great mounds of garbage, putrid rotting fish heads, and clouds of buzzing flies, we got out of the truck, all holding our breath.

"Look, can't I haul the porpoise back down to the beach and throw it back? This is a terrible ending, to bury it under all this filth. Let something in the sea recycle it."

The patrolman shook his head and cleared his throat. "Jack, I don't like this any better than you do. If it were up to me, I'd let you have this porpoise, I know you'd do something good with it. But it's not . . . I can't go against the law. I have to do what they tell me. The Department of Natural Resources isn't making this law, we're forced to enforce it for the National Marine Fisheries Service. The federal statutes say the animal *must* be disposed of by the health department."

The officer helped us pull the porpoise off the truck and lay it on the raw dirt. As he motioned the bulldozer operator to come forward, we got into the truck and drove off. I felt wretchedly guilty about the whole thing. I had interfered with nature and I had caused a mess and now I was sick with anger and frustration. I don't think I ever had a more bitter, frustrating day.

**15**

# The Coming
# of the Horseshoe Crabs

At long last the cold harsh winter with all its frustrations and hardships was over. Even though temperatures occasionally dipped down to freezing at night, covering the bushes and grasses with delicate frost etchings, the warm morning sun would burn it all away. The water was still quite chilly, but as the sun beat down, vapor rose all over the bay, making it look like a gigantic steambath. Great banks of fog rolled in from the warming sea and in the mornings and evenings the land was often cast in mist.

There was a vibrant rebirth of the sea taking place before my eyes as I sat on those Styrofoam boat stalls. I could almost see the hydroids growing, their bright new pink plumes springing up from the pilings, bearing their jewellike orange reproductive buds like tiny bits of glowing coal. The hydroids simply radiated newness.

What a contrast they made from the old wornout strings of mud-covered hydroids that had managed to survive the gloomy winter, grown over with mats of algae and encrusted with bryozoans and sponges. Everywhere I looked over the surface, I seemed to see newly hatched fish fry, so tiny that they could barely be seen against the water background, but nevertheless making their presence known to me.

All over the hydroids, clinging to the lumps of greenish sponges and the potatolike sea squirts, were grass shrimp and zebra shrimp. There were thousands of them, so abundant that when I made a single swoop of my dip net I caught a hundred and five in a few seconds. Over the years I had learned that this was the time, late March, when grass shrimp were at their peak. For just a little while—a few weeks, a month or two at the most—the waters belonged to them. The water was still too cold for predators, and in their absence the shrimp bloomed and bloomed.

The sweet-bay and hickory trees in the hardwood swamps around the bay were beginning to bud out into new green. The marsh grasses that had seemed so bleak after they had turned golden in the fall and drab brown in the cold winter were suddenly turning green and sprouting with newness. The great flocks of mallards, teal, and redheads that darkened the winter skies when they rose honking from the water were starting to thin out and go north. And frequently we saw flocks of Canadian geese winging their way across the bay, getting in their last bit of sea grass and forage on the tide flats before they started on their long journey across the skies.

The water temperature out in the open sea was now sixty-eight degrees. It had risen almost ten degrees in the past two weeks, and we knew that anytime now the great breeding migrations of horseshoe crabs would begin. Then we could replenish our badly diminished supplies and restock our tanks. We had always done a brisk business with horseshoe crabs, shipping them to laboratories all over the world, but this year we had sold more than ever. Researchers at Harvard, Johns Hopkins, Woods Hole, and many other institutions had discovered that *Limulus* blood was a very sensitive indicator of endotoxins, which are poisonous dead cell walls and the residue of bacteria. There was an excellent chance that someday this ancient horseshoe crab could serve as a sensitive diagnostic tool. Potentially, *Limulus* lysate could be used in every hospital and pharmaceutical laboratory in the world, if only the bugs were worked out.

Some batches of horseshoe crabs collected during the summer produced very sensitive blood that clotted immediately when the slightest bit of endotoxin was introduced. Then there were other batches of crabs that were totally insensitive in many instances. It was a long way from being a reliable test, so the government withheld its permission to use it as an official diagnostic tool.

Anne had been intrigued by these differences in the blood of this "living fossil" that has existed for more than two hundred million years on this planet without any significant change. She decided to do her doctoral dissertation on *Limulus*, at Florida State University, and study their natural history. It wasn't an easy task.

If you go to the library and look up *Limulus polyphemus* in the scientific literature, you will be overwhelmed at the volumes of information that have been compiled on this ancient creature's biochemistry and physiology. It is perhaps one of the best-studied animals in the world, yet, despite all of the experiments ever performed on *Limulus*, you can barely find three or four papers dealing with their natural history and behavior. Most of that literature was written by naturalists in the nineteenth century.

Only by learning something about the basic biology of an animal can it be managed as a resource without endangering its survival on this planet. Perhaps after she had learned something about the strange movements and migrations of horseshoe crabs, what they ate, where they lived after they finished laying their eggs, then she might find some clues that would explain the tremendous variation in the sensitivity of their blood. What value did this sensitivity to endotoxins have for the surviving fossil? No one knew.

We did know that any day now the breeding migrations would begin, so every afternoon at high tide we would walk the beach, watching for crabs, and find none. There were plenty of cast shells thrown up on the beach where the crabs had crawled out of their old shells and had grown new ones, but of viable living horseshoe crabs there were none.

Day after day we drove down to the beach, only to see the empty waves lapping on the shore, casting up rafts of floating sea grass and occasional sponges or pieces of sea pork. Every day it was warming up, and sooner or later the crabs would stir out of their wintry mud bottoms and deep water and head for shore to lay their eggs. We knew that any day now there would be the right combination of a warming sea and the proper spring tide, and it would happen just like the oysters in *Alice in Wonderland*:

And Thick and Fast they came at last,
And more, and more, and more—
All hopping through the frothy waves,
And scrambling to the shore . . .

Springtime was pushing on, the days were growing longer, the seas were getting warmer, and still there were no horseshoe crabs on the beaches. We had paid out several hundred dollars to shrimpers for crabs, sold them all, and were getting ready to reorder when we decided to check the beach once again. Somehow I had a feeling that they would be there, and as we drove down the narrow, winding paved road that led to a small sandy berm called Mashes Sands, my instincts were growing stronger. I looked at the tidal creeks that were swollen by the tide. Only the very tips of the green cord grass were above the water, and the wind blew strong out of the south, causing the needlerush and the dark green high tide bushes to sway. When Anne and I arrived at the end of the road where the beach was, we could see them at a glance. The entire beach was covered with horseshoe crabs. There were hundreds of them.

There is no coined word to describe this mass migration of crabs to the beach. Archie Carr, the renowned turtle specialist, uses the Spanish word *arribada*, meaning "arrival," to describe the frantic nesting effort of thousands of ridley sea turtles scrabbling out of the water and up onto the Mexican beaches with excited vigor, laying their eggs, and crawling back to sea. No one ever refers to a *Limulus* arribada, but that was exactly what we were seeing. There were crabs everywhere.

Long ago I had learned not to be greedy during a *Limulus* arribada, although it is an easy thing to do. We had a large building with a huge concrete tank especially designed for keeping three hundred horseshoe crabs at a time. If we crowded more than that into the tanks, many would die. It was best to search the beach and select only the crabs that we needed, and let the rest go.

"Don't touch any until I count them," said Anne, taking her pencil and notebook out of the glove compartment. "When I get finished, we'll collect what we need. You really ought to leave this beach alone and not bother my study area."

I laughed. "Are you kidding? We've been collecting them from here for the past ten years."

"You should collect them someplace else. This area should be preserved."

I followed her, spreading out my plastic garbage cans in strategic locations until she finished counting. She was counting individual crabs, mated pairs, and "pods." A pod was a female that was buried down in the sand with two or more males covering her abdomen, trying to push each other out of the way to fertilize her eggs.

The tide was coming in, and with each wave that broke on the beach, more horseshoe crabs appeared. They arrived in copulating pairs, with large females towing one or two males, or as individual males that roamed up and down the littoral, looking for a female to hang on to. The big cow crabs would crawl out of the water and dig down into the sand, using their heavy shields to plow the ground up like a bulldozer, thrusting themselves forward. The males crawled about frantically. They shoved and crammed themselves into place, each competing for the lady's abdomen. Some of the large females had old mating scars on their shells where the males had dug in and held on. What a sight this frantic orgy of crabs was!

Anne walked hurriedly along, pointing her finger, counting, and scratching lines in her notebook. We walked the entire length of the beach and saw a total of two hundred seventy-eight crabs. It was late in the afternoon when the crabs came ashore, and the usual afternoon fog was rolling in. As we walked over the beach, the wind was pushing more and more of a white swirling mist, and even downshore where there were piles of crabs still copulating, we could see only their dim outlines.

But to us the fog was a blessing, because had it been a warm day, especially on a weekend, it would have been a nightmare of stepping over people to get the crabs we needed. Now the beaches were ours, and they would be until the weather turned sunny. We walked slowly, looking at the crabs, deciding which ones to take and which to let go. I needed fifty males with clear eyes, and twenty-five females, but we had imposed conservation practices on ourselves. Only when the tide changed and the crabs had finished laying their eggs and started to leave would we take any.

The crabs remained on the beach for more than an hour, laying their eggs. The big females dug far down into the sand until only the tops of their spiny triangular abdomens were left exposed, while the males hung on with their well-developed nippers, trying to fertilize the eggs. Hidden from view of any predators, the female would begin exuding her bright blue eggs into the sand, and the eager male would spew forth his milky white semen. We weren't the only ones waiting for the crabs to finish laying. The shoreline was massed with hundreds of sandpipers, the odd-looking little shore birds with long pointed bills that screamed their delight as they hopped over the spawning couples.

The waves washed up on the beach in slow rolling motions, leaving behind bits of manatee grass and brown *Saragassum* algae. At last there were visible signs that the tide was falling. The waves no longer rose high enough to wash the seaweed about. It remained stranded high and dry an inch or two above the water level. Suddenly the huge females began thrusting their way out of the sand, crawling in a semicircle and then lurching into the surf, dragging the male or males with them. It happened suddenly, almost as if the crabs had been given some distant command that beeped into their primitive brains.

Then the clapper rails, willets, and sandpipers went wild. Their cries of excitement were almost deafening as they raced around behind the retreating crabs, stabbing their long pointed bills into the sand, gorging themselves on *Limulus* caviar. They had to make the most of it, because soon the scent would be gone and the sand would pack, and then only by accident would they be able to probe into a nest. Now the waves washing up on the beach were filled with fish. There were tens of thousands of killifish wriggling down into the lower nests, now and then uncovering one and madly devouring the eggs in a veritable feeding frenzy. Schools of catfish had been seen following the crabs that came to the breeding beaches, and they too pushed down into the sand to get their share.

The eggs themselves were a marvel. Unlike the eggs of other marine invertebrates, which are often gooey and tiny, *Limulus* eggs are large—almost a twelfth of an inch in diameter —and are dry. Each female lays hundreds of eggs, and they remain buried about six inches deep. They are washed by the tides and aerated, and sun-warmed in the dry sand when the water recedes. In 1870 the Reverend S. Lockwood, who was intrigued by the eggs of horseshoe crabs, wrote an account of keeping them, which was published in the *American Naturalist*.

He found the eggs to be almost indestructible. He had kept a cluster in some jars over a New Jersey winter, and even though the temperatures dipped down to freezing, a large number of un-hatched eggs survived on the bottom of the jars. Some of the eggs were well advanced, and he described seeing the young crabs revolving around inside their transparent egg shells. This led him to speculate that this might be one reason that the crabs had been around so successfully for so many millions of years. We too had raised them in captivity and watched them break out of their egg casings. The tiny trilobite larvae, so called because they strongly resembled the now-extinct trilobites, buried down into the sand.

Right after hatching they still more or less resembled horseshoe crabs, but they totally lacked a tail, which according to the Reverend Lockwood caused the crabs no end of trouble. "*The want of an articulated tail,*" wrote the naturalist, "*was soon apparent in the case of our little* Limulus. *The slightest obstacle turns it on its back, and not having this organ, which the adult uses so effectively in an emergency, the little thing begins a vigorous flapping of its branchial plates. This causes it to rise in the water; then by ceasing the agitation it at once descends, with a chance of alighting right side up.*"

He went on to describe how one of the larvae molted: "*A few minutes sufficed for it to withdraw itself from its baby suit. . . . At last it came, a person of distinction possessing the articulated rapier. It is a true* Limulus *now, and fully entitled to carry for life the sword of honor, which has ever been the family mark of rank.*"

The indications were good that horseshoe crabs could be bred in captivity, and I was somewhat optimistic about their prospects in mariculture. *Limulus polyphemus* didn't have the complicated life cycle that lobsters and crabs had, with their seven or eight metamorphic stages and prolonged development periods. The mother *Limulus* simply laid her eggs in the sand, and when conditions were right, with plenty of warm sunshine and good rinsing, they hatched and were on their way.

There was so much to learn about this ancient creature that we hardly knew where to start. Interestingly enough, the only other creature that comes high onto the beach to lay its eggs and then returns to the sea is the sea turtle. Archie Carr at the University of Florida has been studying sea turtles for twenty years, trying to find out how they locate their nesting beaches, and the problem is still far from solved.

But the work on horseshoe crabs was just beginning. We knew approximately when they came to shore and laid their eggs, and almost nothing else. Within an hour or two after they arrived on the breeding beaches, they turned and disappeared, leaving an unnerving emptiness behind. If we hadn't arrived in time to see them, we would never have known they were there, save for one or two weaker crabs that had been flipped over by the waves and lay on their backs, flexing their abdomens and forlornly waving their gill books. Always we picked them up by their pointed tails and hurled them into the falling surf to rescue them from their inevitable fate of being picked to pieces by the ravenous sea gulls that patrolled the beaches like buzzards. The crabs are armored tanks except for their gill books, which are sheets of soft respiratory membranes. The crabs also use their gill books in movement, as they race along the sea bottom in the half-walking, half-swimming gait. Sea gulls can easily shred the soft gills with their sharp beaks, and often we arrived on the beach fifteen minutes after the tide changed, and found only empty, picked-clean hulls.

We would often look out over the horizon at the open sea and wonder where all the crabs came from. What clues told them where the shoreline was, and how did they always manage to zero in on the few remote stretches of sandy beaches that are surrounded by a continuous marshy coastline? Year after year they came to the same beaches to breed. Even though there were miles of open sandy beach on Alligator Point and other areas along the coast, only a few stretches of shoreline were acceptable to them as breeding beaches.

During the breeding season, Anne patrolled the beaches, counting crabs that came to shore. Whenever she could get a crew of students together, she tagged crabs, stabbing little plastic tags into their shells that said: RETURN TO FLORIDA STATE UNIVERSITY. On some days the beaches were literally smothered with horseshoe crabs, and on one day twenty-four hundred were counted on a half-mile stretch of beach. On other days only two hundred might be seen, and there were still other days when only a few dozen appeared. Always she arrived on the beach an hour before high tide, waited, and watched and recorded the crabs as they came out of the surf.

The difference in the number of crabs that appeared from one day to the next seemed to be related to wave surge, she learned. If the sea was calm and the water rolled up on the beach in calm glassy sheets, only a few crabs, if any, would appear. If the sea was moderately choppy and the waves washed in on the shore, there would be more crabs. The real horseshoe-crab arribadas came, however, when the winds gusted hard out of the south, the sky was overcast, and the waves crashed down on the sands with force and vigor and lots of white foam. Far out in the bay, the gray waters were whitecapped, and the wind made the sugary white dry sands on the beach swirl and blast in your face.

Then the crabs came, hundreds and hundreds of them, the sea breaking over their shiny brown shells. There were crabs loaded with white barnacles, crabs covered with slipper limpets, crabs with short stubbly growths of hydroids on them, and some with pink clumps of coral and worm tubes on their shells. These were obviously crabs that had lain dormant in the mud, somewhere far from shore, long enough for fouling organisms to settle, and when the right signal was received they stirred from their wintry sleep and started to move to shore.

We had no idea what signal it was that called them, but it was obvious that they were responding to wave surge when they headed for the beaches. Surge, the back-and-forth movement of water beneath the waves, was felt by the crabs, and they followed it to the beach. When Anne attached buoys to the crabs, with plenty of line, she noted that almost invariably they crawled off in the direction of the wave chop. From the shoreline we could see the buoys bobbing along as the crabs moved over the bottom, all together heading into waves. So wave surge effectively guided crabs to and from the beach.

Her studies had not really solved anything about their strange migrations, but just raised further questions that still remain unanswered, because when they crawled over the horizon, they disappeared and were never seen again. One evening we tagged a hundred breeding crabs and watched them return to the sea, dragging their buoys behind them. The buoys bobbed like a sea of crab corks. We watched them until dark and then went home.

The next morning we returned to the beach by boat, all prepared to take compass bearings to plot their movements from the beach. Anne had hoped to see if they were heading in a definite direction or pattern, but not a single crab was to be seen anywhere. We found a number of buoys that had tangled with crab traps near shore and broken off, but all the rest had disappeared. How far could a cumbersome horseshoe crab move? we asked. Surely we could run them down in the speed boat and see where they went, but we ran and ran all over the bay, offshore for two and three miles, in every direction, searching the water for their buoys, but we never found them. No one ever recovered them, either, and no one ever sent back tag returns on the particular lot of crabs.

During the height of the breeding season Anne learned that some of the male crabs returned to the same beach to breed several times. And during the tourist season when the sun was shining and curious people covered the beaches, she would get calls from tag finders asking for their reward. Unfortunately for horseshoe crabs, they have one thing in common with people: they are attracted to pleasant sandy beaches.

When the tides were right and there was enough surge in the sea, the invasion of horseshoe crabs started. Children began to holler, "Stingray! Stingray!" and some got busy bashing the crabs with sticks or throwing rocks at them. There were always one or two people who came down to the sea carrying a gig, hoping to spear blue crabs to take back, and they thrust their spears into the backs of the large females. Sometimes all the stomping feet and splashing people turned the crabs away, but other times the call to breed was so strong that they tried to ignore the mass of humanity and made their way up on the sand and laid their eggs while people stood around gawking and pointing fingers.

"They're *doing* it!" people would say. Young girls would giggle, and jokes would be made, as people watched the male crabs clinging to the females. It was always interesting to hear the diversity of comments when man met crab on the beach. When we picked one up by its tail, and children gathered around to look at its scrabbling legs and flapping gill books and flexing body, someone would always say, "Ugh! Ugly!"

Other questions were. "It won't hurt you?" or, "What good are they?" "Can you eat them?" or, "How do you cook those things?"

They *have to* be good for something. They can't just be horse-shoe crabs, wonderful things all by themselves, a joy to the world just because they exist. They have to serve some kind of purpose for man. After all, the land, the water, the mountains, the air, are all modified and used for the purposes of man.

Man, after all, is the measure of all things.

Man has isolated himself from the rest of the world by his asphalt and concrete and neon signs and lights. And the more removed he becomes from life, the less feeling he has for it. So when a little boy says, "Ugh! Ugly!" he is essentially saying, "This thing and I have nothing in common. It has no place in my world."

It is unfortunate that the child feels that way. He wasn't born believing it. In all probability his mother goes into hysterics when she sees a spider. And God help the world if a snake slithers across her path. Unless this child is pulled aside when he is very young by some naturalist who gives him a snake to hold, and learns that it isn't slimy or unpleasant, he will grow up despising snakes.

As I watched the members of our human race making a scene over the copulating crabs, I couldn't help thinking how far we've been removed from primitive man, who believed that he was one with nature. This separation that we have evolved was especially brought home to me once when I appeared on a television program in Los Angeles with a horseshoe crab and other sea animals in connection with publicity on my book *The Erotic Ocean*.

In the artificial environment of a television studio, with blazing lights, cameras, cables strewn about the floor, and "talent" preening about, I began setting up my aquariums to display the animals I had brought up from Florida. A girl who was a member of the studio staff saw the big cumbersome horseshoe crab that measured more than a foot in diameter, and was kicking its legs with the rhythm of a slow-moving machine, and screamed, "Get that thing away from me! . . . Ugh! It's *alive!*"

"Oh, don't worry about it," I said, "it's not alive. It's made by Mattell."

"Oh, really?" she asked, coming closer. "It *looks* like it's alive. God, it's ugly though. Who would want their child to play with that?"

"Monster toys. It's all the rage nowadays," I replied, trying to took serious. I pressed the crab in its hairy mouth, which is located between its legs, causing it to flex and bend its tail.

The girl moved closer, and I encouraged her to hold it.

"Oh well, if it's made by Mattell . . . " she said, relieved. "Where's its batteries?"

While she was holding it, one of the cameramen walked up and exclaimed in astonishment, "God, that's a horseshoe crab. I haven't seen one since I left New Jersey twenty years ago." Then he saw the crab twist its tail in the girl's hands. "Hey! It's alive, too. How about that!"

"Oh no, it isn't," she said, shaking her pretty head and batting her long eyelashes. She tapped the shell with her turquoise-painted fingernails. "It's a toy. It's made by Mattell."

I winked at the man and he started laughing. "Oh yeah? Where's the key? How do you wind it up? Can't you tell a live crab when you see one?"

By this time I had unpacked a spiny boxfish that was swimming around in a plastic bag, and she saw the uncrated sea turtle on the floor. She looked down at her scrabbling horseshoe crab, dropped it, and resumed her screaming fit. All the cameramen and stage crew who had been standing around were holding their sides howling with laughter, but I had mixed feelings. It was amusing, but at the same time I found the whole business appalling, because there were many people like that girl who were horrified by other living things.

The only link between man and animal that people seem willing to accept these days are Disney films with coy narratives and babbling music, or moronic television shows with lovable porpoises and friendly bears helping cute little children out of some contrived scrape. While people ooh and ah over Flipper, hundreds of thousands of porpoises are drowning in the giant tuna nets of the Americans and having their throats cut by the Japanese. Every year millions of horseshoe crabs are harvested on the Virginia coast and ground into meal to produce chicken feed.

Somehow I cannot imagine a world without horseshoe crabs, or a sea that is empty of whales and dolphins. When sea turtles no longer crawl up on the beaches to lay their eggs because they have all gone into cans of soup, the end of the world cannot be too far off. Empty seas, without the enormous diversity of life, birdless skies, and cultivated land occupied only by people and a handful of creatures bred expressly to service people is something I find too appalling to contemplate.

**16**

# The Sea Grass Beds

All around Dickerson Bay trees and bushes were pushing out their bright new leaves, and spring flowers were popping up in the marshes. The marsh mallow blossomed with lavender, and green shoots of marsh grass pushed up from the rhizomes. During the winter much of the woods and underbrush had burned away by fires set by wardens on the St. Marks National Wildlife Refuge, and through the ashes bright green sprouts were coming up. The flames had consumed the old and the dead, and now the rebirth was taking place.

It was time to restock our tanks with some of this reborn life. The waters around Panacea were once again filled with life. Great schools of speckled trout migrated into the grass beds and streaked into the bays, and happy fishermen were snatching them in one after

another. Word got out that the trout were running off Gulf
Specimen's dock, and people came from all over to fish.

The first little black-tip shark of the season grabbed someone's
bait, and when they reeled it in, we dropped it into one of our floating wire pens and watched it swim round and round with its little
dorsal fin cleaving the surface.

With the sharks were schools of mullet coming inshore, moving
into the marshes at high tide, messily gobbling up the decaying organisms and algae growing on the marsh straw that lay on the
bottom of the creeks. When the tide dropped, and the fish left the
shallow, winding creeks, bunched together in sloughs and congregated around oyster bars for protection, the fishermen were waiting
for them.

At the precise moment when the fish moved into a shallow
vulnerable area just deep enough to run a boat, they would race their
tunnel boats around them, feeding out hundreds of yards of gill nets
from the stern. Before long the fish were trapped in a wall of webbing; some jumped the cork line and swam to safety, but the rest
were hauled in and dumped into the bottom of the boat. There they
fought desperately and furiously, pounding their tails, beating off
their scales, gasping and growing weaker by the minute.

Some of the fishermen unloaded their catch at my dock in ex-
change for a "mess" of mullet when I wanted them. I enjoyed
watching them unload. When they backed their old rusted-out
pickup trucks down to the water, and dumped washtub after washtub of the sleek blue-and-white-bellied fish into their soggy wooden
truck bodies, l was awed and overwhelmed by the vast recycling and
energy-transfer process taking place before my eyes. I watched the
mullet flopping around feebly, gasping away their lives in the alien
world of air, and once again I found myself saying that life was cheap,
easily expended and easily replaced.

In those once-jumping, leaping creatures that moved in vast
schools into the marshes I could see energy. These same delicious
fish that would end up in my frying pan were a gift from the sun. The
sun shone upon the green marshes and penetrated down into the

shallow sea grass beds, and the leaves converted the light energy into sugars and starches through photosynthesis. The plants grew and died and decayed and became part of the food chain. The mullet and other browsers fed upon the decaying marsh straw and sea grass. The trout fed upon the young detritus-eating mullet and shrimp. The little shark in my pen and I both fed upon the trout.

I looked up at the glowing yellow sun in the sky. No wonder the primitive savage worshiped the sun god. He brought warmth and light. I was a pagan at heart.

Down at my feet on the mud flats, fiddler crabs were coming out of their burrows and marching down to the water's edge to pick among the diatoms and other little morsels that were left by the receding tide. More sunlight in action. Now that the water was warming up, Dickerson Bay was becoming cloudy with phytoplankton, untold billions of microscopic plant cells. I watched the crabs from a distance as they methodically picked over the grains of sand. The males had one big claw they could fight with. They waved it at the females to entice them down into their burrows or dueled with potential burrow thieves, and used their other, smaller claw to gather sand and put it into their mouths. The females had the advantage because they had two small claws and used them actively to pick up sand grains, roll them around their feathery mouths, and scrape off the tiny plant cells that had drifted in from the sea.

As the tide dropped, exposing the mud flats and oyster bars around the bay, fiddler crabs came out by the thousands. If they found a dead fish or a blue crab, they would swarm over it and pick it down to the bare bones. Their colors were magnificent: full of browns, pinks, blues, and tans. What made them so impressive was their sheer numbers. They literally covered the flats; in some places almost no mud could be seen, just a sea of crabs. Other times they came out in lesser numbers, only a few roving bands. Herding is a very effective means of protection for fiddler crabs, although it would not seem so. One might think, "Here are so many fiddler crabs. I can get all I want by dashing into the middle and picking them up." But when you actually dash, they scatter, and there are so many that you get confused and are not sure which one to grab first. By the time you collect your wits, they have all dispersed into the marsh and you're left with only one or two.

Strangely enough, the herds aren't disturbed by their normal predators. Maybe when a fox or raccoon or feral hog came down at night to feed upon them, they ran for safety, but I have seen them standing at the edge of the sea, when suddenly the mud erupts and a blue crab dashes out on the land, grabs a fiddler in its claws, crushes it, and retreats to the water to devour it. The herd moves back a few inches and then returns to the water's edge as if nothing had happened.

I delighted in watching clapper rails eating fiddlers. Out from the dense jungle of needle-rush marsh came the funny-looking little birds with their long legs, small round bodies, long pointed bills and their stubby brown feathers. These birds are seldom seen, but when you venture into a marshland you can hear them making a loud raucous warning call when you get too close. When the crabs are out feeding, however, the clapper rails leap out boldly into view and stand there surveying the herd for a few moments, bobbing their long necks in anticipation. The crabs don't pay this avian predator the slightest bit of attention, even when it struts over and snatches one up in its long beak and retreats a few feet to devour its meal.

Only a slight ripple of activity stirs through the margins of the herd. It is as if the crabs were all saying, "Too bad about him, but such a thing could never happen to me!" and they go on about their business of eating diatoms. A few minutes later the long-legged brown bird with the round fat little body makes another raid and again it is ignored.

When the tide starts to come in, the fiddler crabs move up into the marsh. They don't like to get into the water where the other crabs and fish lurk, they always stay at the very edge where the waves lap up. As the water rises they advance slowly up the beach, closer and closer to their protective marsh and burrows. Before the tide actually covers their burrows, they scrape together some sand and make an entrance plug, and when the tide is in they are safely hidden from foraging fish.

The constant digging up of the soil permits the water to percolate down through the marsh, aerating the soil and improving the drainage of the marsh. Fiddler crabs have a tremendous importance in converting the energy of the marsh grass into a usable form, providing food for other creatures directly as prey, or indirectly by producing their copious fecal pellets. Often I have walked on sandy salt flats, which are usually covered by an inch or two of water, and watched the crab pellets being carried away by the sheet of water. Hordes of infantile mullet and killies so small that you can hardly see them hide among the saltworts and peck and grab away at the rich fiddler pellets.

Fiddler crabs are an important teaching and research tool used to study neurosecretory hormones. Their pigments are located in special cells called chromatophores, which are scattered over their carapaces and legs. These cells expand by day and contract by night, causing the crabs to appear much lighter when the sun goes down. This rhythm persists under conditions of constant light or darkness. The size of the chromatophores also changes with the tides. Physiologists have found that this rhythm is controlled by hormones in the crab's eyestalks, and they can remove the eyes and cause the crab to lighten. By making an extract of the eyestalks and injecting it into the fiddler crab, the color changes will be reversed. This is only one of many experiments that have been conducted with this crab. Biologists at Woods Hole's Marine Biological Laboratory have been working with *Uca pugilator* since 1937.

We didn't want to collect too many in one spot, so we often went up and down the north Florida coast rounding up herds of fiddler crabs, driving them like cattle and gathering them up in garbage cans. St. Joe Bay in Gulf County was eighty miles away from Panacea, but it was worth the drive. On almost any warm day we could see acres and acres of tide flats covered with fiddler crabs.

On these same expansive flats that stretched far out into the bay there were also hordes of juvenile horseshoe crabs. Now that the sun was warming up the tide flats and the waters were beginning to lose their chill, juvenile *Limulus* were starting to emerge from deep in the sand. During the winter the flats had been empty of crabs, but it was May, and the sand was crisscrossed with hundreds of horseshoe-crab trails, running in all directions, twisting, winding, turning and returning. Anne would take one part of the tide flat to watch, while the rest of us would work another and collect. The little crabs could never be found at high tide. When the water covered the flats they simply dug down into the sand and disappeared. They conducted most of their activity when the tide was receding.

When the tide had finished falling and all the water was drained off the flats, the tiny horseshoe crabs ceased their activity once again and hid down in the sand in the midst of their matrix of trails that seemed to have no beginning and no end. How great it was to walk barefoot over tide flats and look for them, free of our long-uncomfortable boots. But even with the extra comfort of pleasantly warm weather, finding the crabs wasn't any easier when they weren't moving about and clearly in evidence.

There before us on the ribbed sand was a bewildering complex of trails. Man, being a logical animal, will look at the trails and say, "Somehow, if I follow this trail long enough, I will come to the end and there will be a horseshoe crab." But the trails twist and turn, stop, begin, end, and twist some more. If you were to stretch out one of these little trails, you would find that each crab travels more than a hundred feet in the course of a day.

We had learned to totally ignore the trails if we wished to retain our sense of reason and find any crabs. We walked the flats, glancing over the sandy puddles, looking for movement. Suddenly we would see a little clump of sand stirring along the bottom in the midst of a trail and we knew we had our crab. Usually the abdomen or just a portion of the tall protruded at the end of the trail.

Although we could find juvenile horseshoe crabs on sand flats closer to home, St. Joe abounded with them as well as with fiddlers. And while it was worth the eighty-mile drive to get *Limulus* alone, there were other things there that made it even more worthwhile. Among the crystal-clear grass flats we could snorkel along the shallows, filling our diving bags with every imaginable creature ranging from brown-shelled horse conchs, with their great scarlet meaty feet protruding from their shells, to hermit crabs wearing colorful pink anemones on their backs. No other bay in north Florida had the diversity of St. Joe Bay, and that was truly ironic. At the mouth of the bay stood a huge industrial complex with chemical factories and paper mills belching smoke and pollution into the air and water. If the currents had been different, the bay would have been a lifeless polluted mess, just the way it was around the outfall of the factories. But the currents carried the sulfide wastes, humic acids, bark, wood chips, and chemicals offshore, and the head of the bay was beautiful and green and full of life.

Even while we walked the tide flats, we were partaking of the diversity. Mixed in with our buckets of little tan horseshoe crabs were turbin snails with their pearly opercula, and soft-frilled sea hares, *Bursatella leachiplei*, which were all over the flats. There was so much life on the tide flats that Leon decided not to go diving with us. "You go ahead," he said as we started toward shore. "I'll stay here and build up on our stock of horseshoe crabs. Besides, I want to dig some *Cerianthus* anemones—we're just about out of them, and they're all over the place here."

St. Joe Bay had always been good to us. I could hardly remember a trip there when we hadn't come back with our buckets crammed with life. All we had to do was squeeze into our wetsuits, put on our masks, fins, and snorkels, take the boat out on a particular rich patch of waving submarine meadows of turtle grass, and dive overboard. That first shock as the cold water pours into the spaces of your wetsuit is never pleasant, but it only lasts a second. Suddenly you're in a world of endless grass and beauty, in the midst of an experience that there is no getting used to. We had made hundreds of trips to St. Joe Bay, and each time it was magnificent.

On a collecting trip you go as a predator, not as an observer. While roving the sea bottom like a shark looking for prey, you don't have time to really observe animals or watch them feeding or see how they interact, how they mate. You have to select and then move on, diving down and tearing up a black tunicate with golden star-shaped zooids and putting it in your diving bag, and then swimming on, looking, looking, looking. You may come upon two ugly brown spider crabs tearing apart the flesh of a dead conch, or you may arrive upon the scene of a huge red-footed horse conch devouring a left-handed whelk. For a moment you may watch—and marvel at how the conch expands its huge orange flesh around its prey and smothers it—and then move on. And everywhere surrounding the battle there will be dozens of hermit crabs, attracted by the smell of death, waiting, hoping to get the shell.

But these are only flash observations. Your real mission is to fill your bag with as many different creatures as you can get. And our mission that spring day was octopuses, clams, conchs, sea cucumbers, tunicates, sponges, and so on. A scallop snapped its valves shut as I passed overhead. My eye caught the quick movement, and I dived and got it. A few yards farther, another began to swim off the bottom, rising up from its grass cover by clacking its valves together, opening and shutting them and moving around rapidly in fast jerky hops, and almost escaped into the foliage as I caught it.

What a glorious bay this is, I thought, watching a school of pinfish browsing among the turtle grass. What a tragedy to lose it, when we had come so close to saving it forever. The State of Florida had considered buying up all the flatwoods, marshes, and dunes surrounding the bay under its endangered-lands acquisition program. The traditional battle ensued, environmentalists and a few commercial fishermen on one side, businessmen and resort interests on the other. In the end, the proposal was shouted down in a public hearing by rural north Florida rednecks who wanted to see the area developed.

When we launched our boat at the fish camp, the owner had gloomily told me that a friend of his had just gotten a contract to bulldoze a hundred acres of trees at the head of this pristine bay. There were plans to build condominiums on the dunes across the bay, swimming pools, a tennis court, six hundred units . . . imagine, thirty thousand people out there. How long would St. Joe Bay be so pristine? As I kicked my flippers and glided over the shallow grass beds, I was taken by the beauty of all the pink and white sea urchins, *Lytechinus variegatus*, that were clustered all over the bottom along with the brilliant orange clumps of branching bryozoans, so much like a flower garden. This bay really was priceless.

I dived down five feet to the bottom, grabbed an urchin, and cracked its shell against a heavy conch shell. When the urchin broke apart I washed out its brown insides, leaving the bands of well-developed gonads attached to the inside of its spiny test. They were just what I was looking for. The eggs were getting ripe and in a few more weeks *Lytechinus* would be ready to harvest. I dropped the urchin, and instantly two pinfish appeared from the grass beds and began greedily tearing and shredding its soft parts.

Soon everything in the sea would be spawning, a whole ocean filled with sperm and eggs and larvae . . . productivity, fecundity . . . it was great. I passed a tulip shell laying a big clump of eggs, which came out as translucent fan-shaped capsules that the snail cemented to an algae-grown clam shell. I looked at it, and could see the tiny pinfish red embryonic snails inside, looking like their parent. Who would object to an ocean full of tulip shells with their pretty banded shells that looked like the whorls of a barber's pole? Probably the venus clams and smaller snails that they fed upon.

Sea urchins are good and useful creatures and serve a valuable function when they are in balance with their environment, but when they go out of balance they can damage the grass bottoms. As I swam on, I came across great cuts and slashes in the grass beds where motorboats had run aground at low tides and had dredged their way out to deeper water, their propellers ripping and tearing rhizomes as they went. Some of the swaths had been made years ago and were now filled with soft oozy mud. Turtle grasses grow very slowly. Sometimes they regenerate and sometimes they never do because the bottom erodes away. I came across big gouged-out portions of the grass beds where commercial scallop boats had worked the bottom over. They used heavy dredges and pulled them for hours at a time, leaving big scars and exposed rhizomes behind. The fish-camp owner and the sportsmen were talking about getting a law in St. Joe Bay to prohibit scalloping with dredges, because they knew that once the grass went, so did their fishing.

Turtle grass nourishes the ocean, providing tons upon tons of detritus and plant material. The grasses grow and flourish during the summer and then start to die back in the fall. The waves and wind tear up the old grasses and dump them on the shore in great rafts, where they dry and decompose in the sun and wind. The grass is broken down even more by the abrasion of sand on the beach, and then it washes back out to sea, where it is further decomposed by bacteria and ends up providing food for almost every imaginable creature.

Only sea urchins and green turtles feed directly upon the grass. Since green sea turtles are rather rare in the Gulf, it is the urchins that keep the grass cropped down. When the meadows are too thick because there is too much silt trapped among the leaves, nothing likes to live in it. The grass has most life in it when it is only about half grown, in the spring and in the late fall. In the winter it dies down to nubs of green sticking up from the sand. There are all these balances in nature . . . one endless chain of checks and balances . . . there is so much to learn, so much to understand.

Before long I had accumulated a fine assortment of scallops, orange starfish, sea squirts, tadpole cling fish, and other creatures, but I had yet to see my first octopus. I hoped Anne and Doug were having better luck than I was, and just as I was beginning to worry if we were going to fill our order, I spotted one.

What I actually saw was not, as one might expect, a softbodied cephalopod stretching its eight long arms out, grasping the bottom with its suction cups and pulling itself along. Nor did I see it jetting itself through the water with all eight legs bunched together, giving the animal a streamlined teardrop shape. What I saw was merely an encrusted sunray venus clam lying on the sand. But that was enough to tell me that there was probably a dwarf octopus inside.

*Octopus joubini* is probably the most shy of all octopuses, and probably the smallest one also. Although they often hide in crevices in rocks, we frequently found them in the shallow grass beds hiding in gastropod and bivalve shells. When we saw a whelk shell lying on its side with a bunch of broken shells pulled tightly against the mouth of the shell, forming an artificial operculum, then we knew an octopus was inside. We also knew it when we saw a sunray venus clam or a cockle, which is normally partially or completely buried down in the sand, sitting blatantly on the surface in the midst of a grass bed. The octopus inside used its powerful little arms to keep the shell closed, and if we tried to pry the shell apart, the little beast did its best to force the valves closed. It would blush red and become angry, but only if we pried the shell open would it depart, in an explosion of black ink.

I found another octopus in a cockleshell approximately ten feet from the first, and I decided to open the valves. The octopus did its best to keep them closed, but I was stronger, and at last it released its grip, but it wouldn't leave. I could see why, because inside, attached to the bottom of the cockleshell, were about a hundred transparent egg capsules, and I could see that some of them already had tiny developed octopuses inside.

These were especially welcomed because we had customers who wanted to raise them in their aquariums. Some skilled aquarists had succeeded not only in getting the eggs to hatch, but in raising the young to second generations.

*Octopus joubini* do not make the best aquarium specimens, because you seldom see them come out of hiding. But if you watch at night, you can see them emerge from their shells and forage around for food. If you put a hermit crab or small mud crabs in the tank, they will wait until the crustacean comes close enough to their hiding place, then whip out their arms, blushing a reddish brown, and drag the prey inside. Within a few minutes the refuse will be thrown out, bits of legs, shell, and antennae. The brooding females seldom eat. They remain with their eggs at all times, washing them, caring for them, seeing that they are protected. When the babies finally hatch and go their way, the mothers always die.

Suddenly my luck was changing, and I was finding octopuses all over the place. After two hours of swimming around the grass beds, I had six that I was sure of, and two that I wasn't. There were two whelk shells that were so massed over with barnacles and oysters that I couldn't really be sure there were octopuses inside. We would just have to wait until we got back to the lab. Then I would place the shells on a dry table and wait, and if there were octopuses present, they would become uncomfortable from the lack of water and come squirming out of the shells.

I kept looking, but my bag was getting heavier, because I had also picked up several large conch shells that weighed three pounds apiece. I was also getting cold and tired. I knew that I would have to call it quits soon.

I passed a spider crab, grabbed it, considered it for a moment, and dropped it. We could get plenty of spider crabs off our dock in Panacea . . . there was no sense taking this one back . . . besides, they ate too much. If you didn't watch them, they'd eat the tanks barren. I'd walk in some morning and see a one-dollar spider crab sitting there having lunch on a seven-dollar sea robin—it happened all the time.

I came across a pair of horseshoe crabs that were hurrying along the bottom coming from God knows where and going perhaps to the shore to lay their eggs. Why were they always copulating? Here they were several miles from the shore and they were already coupled. I had been aboard shrimp trawlers operating out in the Gulf, ten miles from shore in sixty feet of water, in January . . . and horseshoe crabs would come up with the males clinging to the females. Some of the cows had mating scars on their shells.

That paper Anne had brought back from the library, the Reverend Lockwood's "The Horse Foot Crab"—he wondered about it too. All those juveniles that Leon was collecting, there was no way to tell the males from the females. They had to mature first. As Lockwood put it: "*It* [the male *Limulus] then has a moult, from which it emerges, having received its large claws, or literally its nuptial hands. What change there may be on the emotional side who can tell when master* Limulus *assumes the* toga virilis *and is old enough to 'propose.' This may be asserted of these very decorous and monogamous people, that among them premature marriages are unknown, for however soon the lady may be ready to give her heart, not until maturity of age can the gentleman possibly extend to her his hand.*"

Why can't scientists write like that nowadays? I thought as I watched the horseshoe crabs disappearing into the grass beds. If Lockwood were alive today, he'd have to write in dull tedious scientific-ese if he wanted to get published. Maybe that's what I am . . . a reincarnated nineteenth-century naturalist. . . .

My mask was fogging up. I lifted it and felt the cold water run in against my nose and eyes. I blew the water out. I could see again. But it was getting near time to quit.

I looked up and saw Anne climbing aboard the boat while Doug hauled her bag out. I wasn't the only one who was tired. A few minutes later we sat in the little boat, sorting our octopuses into separate buckets, separating out the heavy conchs from the red-beard sponges and the fragile arrow crabs. Doug had found a sea squirt full of bright orange flatworms, and Anne had picked up two big sea horses. St. Joe Bay had really been good to us that day.

As we sat there working, the silence of the morning was broken. From across the other side of the bay we heard a rasping, buzzing sound and then a soft crash. It was a sickening, all-too-familiar sound that shattered our victorious, joyful mood.

It was the sound of a tree saw.

# 17      The Hurricane

It was well into June and the waters had been calm and beautiful for the past few weeks. In fact, they had been too calm, and each day at high tide we drove to the beaches at Mashes Sands only to see the waves gently lapping up on the shore. There wasn't enough surge on the beach to bring in more than two or three pairs of horseshoe crabs. Each day I hoped the weather would get rough, that it would rain and the winds would blow hard from the south, turning the seas rough and choppy.

Never would I have hoped for that if we hadn't been in such desperate need for horseshoe crabs, but a large, large order had come through. We needed two hundred big hulking crabs for a biochemical laboratory, and these people promised lots of further business and our finances were looking grim as summer approached. The seas remained calm and glassy for all our hoping, and despite all our efforts of traveling around to offshore sandbars where horse-shoe crabs often congregated and diving in St. Joe Bay, we had come up with less than half the quota.

I listened to the weather forecast in the morning over my cup of coffee and heard with some mixed feelings that an early large hurricane was brewing off Cuba and had moved into the Gulf of Mexico, with center winds of only seventy-five miles per hour and a forward speed of five or six miles per hour. They gave the latitude and longitude and advised their viewers to keep watch over the next few days. A storm like that could go anywhere; it's the kind that you hope will hit someplace else, like maybe Louisiana or south Florida, but won't come into your backyard.

Even though the news reported bad weather out in the Gulf, there was no sign of it. Outside, the sky was as blue as could be, and the sea was still as calm as the proverbial millpond. There was a strange look to the sky, however, because all the huge boilerlike summer clouds that stretched high into the sky, the cumulus that towered up above everything, were gone. In their place were small-er, scattered little puffs, all broken, and yet there was a strange regularity to their pattern.

It was the following morning that we began to really see signs that the Gulf was disturbed. The south winds came, all right—winds began to whip up, and suddenly the waters in the Gulf became turbulent, gray and frothing with whitecaps. We waited until high tide, then Leon, Doug, Edward, and I gathered up all our plastic garbage cans, loaded the truck, and headed for the beach, expecting to see hordes of horseshoe crabs converging on the shore.

We looked over the same breeding beach they had used before, every time there was rough weather, but there wasn't even one crab coming out of the waves. In past years when the weather had really been violent, the crabs stayed away from the beaches. We walked the shore, looking, hoping, but only great piles of floating sea grass were being cast up.

Leon looked out over the whitecapped horizon and shook his head. "Those crabs know there's a damn hurricane coming. They ain't crazy. If that thing hits in here it would wash every one of their eggs out and maybe kill the lot of them."

"Yes, but how do they know?" I wondered aloud.

"Hell, that's just nature," Edward replied, trying to light his cigarette in the wind, sheltering the match with his hand.

"I sure don't like the way those clouds are moving," Leon said, pacing nervously around the beach. "That damn hurricane might come right in on us. Where's it at now?"

"I heard it was two hundred miles off Alabama," Edward replied. "Maybe we ought to start getting the boats out of the water and tying the floating docks down."

I looked up at the sky, and there above the swirling gray clouds, making low flying circles, was a strange-looking black bird. It seemed ominous and dark and it beat its wings slowly, flying with the wind. "What is that?" I asked, pointing.

Doug, who was something of a birdwatcher, squinted up at the sky. "That's a frigate bird. I've never seen one this far north. They live further south. I think Cedar Key is as far north as they get."

"I don't like the looks of him," said Edward. "I wish that son of a bitch would fly someplace else. Let's get out of here."

We worked through the rest of the day, having forgotten completely about the horseshoe crabs, and made our boats fast. We ran *Penaeus* far up the Ochlockonee River, where it would be protected, and hauled our smaller boats out on trailers, bringing them up the street to the laboratory building. Mary Ellen and Anne worked in the office, getting together the important files, correspondence, manuscripts—the whole business, just in case we had to evacuate. I didn't think we'd have to, but there was no need to take chances. Several years ago we had learned our lesson when the huge storm Camille came ashore in Mississippi, killing hundreds of people, wiping towns off the map, tearing shrimp boats to pieces.

A number of shrimpers who lived in Panacea had been through that storm there. They had lost their boats and they told horrible stories of riding their vessels through downtown Pascagoula, of leaping ship and hanging to the railings of an overhead bridge, only to watch their sixty-five-foot boats smash to pieces. Those who managed to ride out the storm still had harrowing memories haunting them weeks later of pulling up bodies at sea. After that storm, everyone was calling the Red Cross in Mississippi, trying to locate this shrimper or that deck hand who had been working over there.

Yes, Panacea remembered. It could happen here. The little swampy low-lying community could be flooded out. In the cafe, old-timers talked about the great storm of thirty years ago that drowned four men because they were unable to escape over the washed-out roads and the rising waters. There was a lot of talk throughout the evening, a lot of bustling activity, and then everyone turned in to watch their televisions and see where the storm was. Maybe it would turn around, maybe it would go elsewhere . . . all you could do was hope.

"Evacuate, evacuate immediately," a metallic voice blared in the darkness at two o'clock in the morning. "This is the highway patrol. Leave your homes immediately before the highway is flooded. The hurricane is headed this way. Evacuate immediately!"

I jolted awake from my catnap. After one o'clock in the morning we had finally decided to turn in briefly, as over and over and over the weather forecaster gave the position of the storm. It was going to hit somewhere between Mobile, Alabama, and Cedar Keys, Florida, but at that time the evacuation order had not yet been officially given.

I looked out on the street and it was bustling with activity. Headlights were lighting up the road, and the highway-patrol car's flashing blue lights were eerie. Civil Defense trucks were out. I could hear people shouting and working. A horse galloped past, running behind a pickup truck on its way to pasture high up in the woods. The general store and gas station were open and crowded.

I didn't have to be told twice. The wind was gusting in my face, the water was inching up under our house, carrying debris and sea grass with it. The tide was higher than I had ever seen it. In just a few hours it had risen several feet and in the beam of my flashlight I could see that it was covering my dock. Only the tops of the railings were sticking out. "Evacuate immediately. . . . This is the highway patrol. . . . This is your final warning. . . . Evacuate!"

I guess we're intellectuals. We passed the test that night anyway, because instead of packing up our household trappings, such as television sets, furniture, clothing, and dishes, Anne and I frantically stuffed our books into Styrofoam boxes and loaded them onto the truck. Leon and Edward came over to help. There was a great advantage to having a business that kept hundreds of big Styrofoam ice coolers in stock. Instead of horseshoe crabs we filled them with papers, manuscripts, books, and company records, but there was no time to pick and choose. There was a limit to room. The rain began to pour down on us, the winds gusted harder and harder, and every moment the water was coming closer. The wheels of the truck were now standing in water. We pulled the main fuse on the electricity, grabbed the cat, slammed the door, and drove to the laboratory, which was on higher ground.

There we killed the power. All the tanks of bubbling water were suddenly silent. Our pet sea turtles surfaced in hopes of getting fed. Down the street we could hear, "This is the final call," from the Civil Defense. "There is a shelter at the Wakulla High School. If anyone needs a ride please come forward. Evacuate immediately!"

As Anne and I drove down the highway in our truck, behind the long string of cars and pickups, in the headlights we caught a glimpse of the creek. The waters were swollen, and the very tip of the tall needlerush marsh was sticking up between the rolling waves. We could see a boat washing up against the shoulder of the highway. We weren't leaving a moment too soon.

We went to a friend's house in Crawfordville, an inland town about twelve miles away, and spent the night there. Normally I would have been up worrying about what was going to happen, but I was too exhausted. We crashed into bed and slept as only the truly weary can. Leon had gone on to the schoolhouse. Others took refuge in the motels in Tallahassee. It was strange to feel expatriated, dispossessed, and chased out of one's home. People and animals have felt that from the beginning of time when a storm comes in and water claims the land even briefly. It doesn't take long for the sea to come dashing in, to swamp the ground, to flood the banks of the rivers with its salt water, to soak down into the soils and begin the slow death of the trees. Yet it has happened, time and time again, down through the ages over millions and millions of years.

High tide was to come at six that morning. From what we gathered from the television, the direct path of the storm was still hazy, but it was supposed to come ashore east of Apalachicola. Winds were now approaching a hundred miles an hour, and the seas were six feet above normal. Into the dawn the heavens rained down in a solid sheet of water. Sitting in our friends' concrete block house, removed as we were from the sea, we could still feel the full force and violence of the approaching hurricane. As wind-driven water was forced through minuscule cracks in the window, and the roof began to leak, I envisioned the raging seas pounding the shores of Panacea. We turned on the television and my worst suspicions were confirmed: "Houses have been extensively damaged, docks are being swept away from Steinhatchee to Apalachicola," the news broadcaster said. "One eyewitness saw the roof on Alligator Point Marina struck by a tornado. The damage is massive. . . ."

I had no illusions. Last year's tornado that bent and twisted my dock was just a summer breeze compared to this. We sat and waited, hour after hour, while the winds howled. When the eye of the hurricane passed over, the rains quit and died down, and for a little while I contemplated going back to see how bad it really was. But the Civil Defense warned that electrical lines were down all over the place, bridges were still flooded out, and many roads were blocked by fallen trees. The rain and winds began with a vengeance again, and continued until late afternoon the following day. Finally the storm passed over, the winds quit screaming and the rain finally stopped.

It was time to return to Panacea. As we drove back into town, we could see the destruction laid out before us; the water from the bay was still swollen in the roadside ditches for more than a mile outside the city limits. Everywhere there was debris and rubble. Houses were flooded, especially the ones on the bay built close to the ground and not on stilts. A house trailer was standing in water; a stove washed back and forth through the back wall. Old pilings, debris, garbage, and piles of marsh grass had covered the road and washed into people's yards. The storm hadn't been as bad as it could have been. We had a lot to be thankful for. There were still houses standing. Only a few people had really lost their homes. Some people had all their furniture ruined, but most of the damage was inflicted on people who were living too close to the water.

We couldn't drive down to the dock on the back street. It was a journey that had to be made on foot. Piled knee deep was tons upon tons of dead needlerush marsh straw that had been washed out of marshes all over the coast and flung high up on the shore. The woods were buried beneath the aged dead grass, and the highway was completely covered. If you hadn't been to Panacea before and were seeing it for the first time, you would never have known that beneath that rubble from the sea was a paved road. I stood there in shock, looking at it all, and shook my head sadly.

"I can't get over it . . . all this enormous productivity of the oceans is wasted. Look at it, there must be a million tons that will never get back into the estuary."

"You really are a fanatic!" snapped Anne. "Here we are with maybe half the lab washed away, the dock destroyed probably, and all you can worry about is the marsh grass. That is something!"

But for some reason, that was all that seemed important. As we stepped over the soggy grass, we saw grasshoppers, crickets, and spiders jumping and crawling everywhere. They too had been disaster victims. There were black marsh crabs hundreds of feet from the sea. We stepped past the bodies of drowned marsh hens, and I smelled the reek of diesel fuel. There were fuel-soaked birds among the corpses. A huge fuel tank had been washed in and was still oozing oil. As we walked down the street toward the dock, we saw people coming back to their homes. A television had washed out on the highway, and clothing and furniture were strewn about. There were more dead birds, drowned marsh rats were everywhere, drowned legless lizards, a dead possum. How cruel the loss of life.

It was almost a quarter mile to the dock. I passed a baby diamond-back terrapin crawling up into the woods, and I stopped and picked it up. It needed to go back to the sea. Snakes were everywhere. The yellow striped marsh snakes had been washed out of their interface between land and sea and cast up on the shore. And as people returned to their homes I could hear gunshots going off as they met the bewildered, harmless serpents. Everyone on the back street was digging out, raking the piles of marsh straw away from their doorsteps. Great piles of prickly-pear cactuses had also been ripped up from  the offshore islands and carried ashore. The trim little gardens of the Yankee retirees and their neat lawns, just like the ones they'd had in the Chicago suburbs, were smothered with the soggy grasses. Again, the blast of a shotgun, the crack of a pistol as another life was snuffed out. I found another terrapin, sitting there bewildered, as if wondering where the sea had gone. It was all so terrible.

When we arrived at the place where the dock had once stood, I stood there in dumbfounded amazement and disbelief. The dock was gone. It wasn't there. The bay looked strangely open and free and wild where the dock had been ripped up by the roots, broken into sections, and hurled up into the marsh. What a scene of ruin and destruction! It was incredible. It looked worse than anything I had seen so far. Its pilings protruded up from the marsh grass in a grotesque twisted fashion, like a running deer that had been brought down by a rifle. The floating docks were broken into sections, but the sections were all still intact. They had floated off their pilings with the rising water and had washed up onto the highway and come to rest on top of the marsh straw.

Still in shock, I walked on up the street toward the laboratory. What was left? I wondered. Was there anything to go back to? The straw beneath my feet was soft and spongy even though there was pavement beneath it. I stepped through soft muck. Black crickets jumped and bounced all over the place. There were millions of them. I wondered, idly, why in all my years of stomping around marshes I hadn't seen so many crickets. Where had they been hiding before the storm dispossessed them?

I had a pleasant surprise when I returned to the laboratory. The chain link fence that encircled our facility had managed to keep out the tons of marsh straw and debris. It was all piled up against the wire. Water had flooded in, but it didn't get into the offices. It had washed through the lab, but everything was intact. The water didn't cover our tanks. All our inventory didn't swim out. I quickly surveyed the damage. There were ruined air-conditioners, minor damage to the buildings, flooded-out lawn mowers, things like that—but nothing major. The roof was still on, and the offices, desks, and filing cabinets were exactly as they had been. Our house on pilings was still intact and relatively undamaged.

I flicked on the power switch, and the tanks started bubbling away. The specimens had been without air for nearly twenty-four hours. A few fish had perished, and an eel protruded from the sand tank, gasping. Oxygen bubbled in. I turned on the sea water line, but nothing happened. That would have been too much to ask for our pump to come through the storm undamaged. I dreaded the nightmare that would be involved in fixing that and repairing the damage.

The telephone rang. "Gulf Specimen Company," I said, out of habit.

"Hello, this is Purdue University. We'd like to place an order."

"Could you call back tomorrow?" I said in a broken voice. "We've just come through a hurricane."

"But this is very important . . . we need the material for class on Wednesday and—"

"I'm sorry," I said. "Please call back tomorrow. We're closed."

The girl on the other end finally seemed to grasp the situation. "Will you be shipping orders this week?"

I looked around at the tanks bubbling away. There were still animals, and we still had boats that were sitting on trailers with motors on them. Why not? "Yes, we will be shipping orders. Please call back tomorrow."

I hung up. Leon walked into the lab and shook his head. "We're ruined. I ain't never seen such a mess in all my life."

"Oh, we're not ruined," said Anne. "We just lost the dock. You had Gulf Specimen Company a long time before you had a dock. There's too many people who depend on us. Maybe we can get a disaster loan. I heard the President just declared north Florida a disaster area."

"I don't know how we can get along without that dock, though," said Leon miserably. "We use it for so many damn things. The animals we grow on it, keeping our boats there and all."

"The animals." I remembered the culture racks, the oyster strings, the whole bit. All the sponges and tunicates, all the living rafts of animals, the whole living dock. "Maybe we can get the floating dock back into the water before everything dies. Maybe it isn't too late."

We hurried back down the street to where the wreckage of my dock lay. Like the various boats in town that had been jerked up from their moorings, some of the floating docks rode the waves up past the marshes, over the woods, across the highway, and into a neighbor's yard. There they were stranded. The tide returned to the bay and they lay there on the ground. I could see the strings of barnacles, the greenish yellow sponges, and the hydroids starting to dry up in the afternoon sun that was peeking out from behind the clouds. There were even a few grass shrimp still alive, but as we looked over the wreckage we knew that the living dock was dying.

"Come on," said Leon, "let's go get the truck and drag it back to the bay. Maybe we can save something. It's better than sitting by and watching it all die."

It was something to do. And in such a hopeless situation, something, some act of trying to save life, was better than nothing at all. While Leon went off to try and get the truck down to the dock, I walked around the broken twisted wreckage of the main dock. People were coming by now to look at the wreckage. There were some who came to gawk and shake their heads and make words of pity. These were the tourists, the people who drove down from Tallahassee and from surrounding areas to take it all in. There were other people there, local people, who had come to help.

With rakes and shovels we cleared a path down to the water, pushing aside pilings and heavy debris so that the truck could tow the floating docks down to the water. We pushed aside the splintered lumber, raked away the ugly jagged nails that could perforate a tire or a foot, and strained to push the great floating docks in a desperate attempt to get them back into the sea. My neighbors and friends appeared, some of the fishermen who didn't quite understand why they were doing it, but they were there to help regardless. Leaving a great heavy trail through the mud and straw, the truck dragged the boat stalls down to the water and we pushed them in. They floated and then Leon and Edward lashed them together like one big barge and anchored them to a lone piling that stood out in the bay, a single piling that had withstood the ravages of the storm.

It was futile and I knew it. Perhaps I would save a few creatures here or there, but what did it really matter? I glanced down at the trail of twitching, dehydrating porcelain crabs that had fallen off the dock as it was pushed along. I found a bewildered blue crab that had taken refuge on the dock as the storm waters came crashing down. There was a stone crab on the ground, a female with orange eggs. I dropped her into the water, hoping that she would survive, but the crab lay there on her back, legs twitching spasmodically. Over and over again the lesson comes home that in nature life is cheap, so very cheap. It doesn't seem to make a whole lot of difference who destroys it, man and his machines or nature. There is a difference, though. Man cannot replace what he has destroyed, but nature can.

I walked around the uprooted, contorted dock pilings, gathering up whatever wiggled or crawled, and hurriedly dumped it back into the bay. There was no saving those millions of creatures, and we knew it. What we were doing was only tokenism, but somehow it was important. I watched the barnacles clicking and bubbling, saw the valves of the oysters beginning to gape on all those pilings. There was nothing I could do for them. Now I felt guilty about it, because somehow I had been responsible for their creation by providing them with a habitat, but I could do nothing about their impending death. The tunicates were drying up and dripped water. The sponges were desiccating. Now and then there would be a live toadfish, wriggling back and forth in the mud almost as if it hoped it could wriggle itself back into the water. But the water was almost a hundred feet away.

There were crab traps still attached to the ropes that were tied to the broken dock railings. They had been jerked up along with the dock and carried to shore, and now they were filled with despairing bubbling blue crabs. We emptied the traps and released the crabs too.

Am I being silly, being compassionate for such lowly creatures as barnacles and shipworms? I wondered. Why does man feel any emotion for this life? Overhead, the sun was shining, the clouds disappeared and it was a beautiful day. Just like that, first a holocaust and then all bright smiles like nothing had happened.

The next day the smell of death was everywhere. While we cleaned up and cleared the road, straightened up what we could, from a block away you could smell the odor of death and rot. To a passer-by it might have been the smell of a fish house that had dumped waste water on the ground or left a box of fish guts out in the sun, but to me and my staff it was the smell of death and horror. The oysters gaped their valves, the barnacles dried up in their closed shells, and flies buzzed around by the billions. In a short while the living dock was reduced to a skeleton, with just the remnants of life bleaching away in the sun, all the calcareous creatures with shells and slimy tubes. There were brown mussels with their byssus threads still anchoring their shells to the pilings, and in places the wiry brown hydroid bushes remained, dry and brittle. When I grabbed them in my hand they crushed into a powder and I sprinkled them into the marsh.

A cool wind blew down from the north. The sky was bright and blue and there was a newness about the marsh, which had been washed green and purged of organic wastes. I looked at the open space where my dock had been and I had to admit that it looked beautiful, uncluttered, wild and open water. Nature had purged herself of man's structures. All up and down the coastline she had cleared away much of her cluttered beaches, she had ripped up docks and wharfs. But she had also chewed away her sand dunes, tearing up the sea oats by their roots. Great towering slash pines that had grown down by the water were dying now from the salt water, their needles turning red. The dwarf sand live oaks were burnt by the salt and their leaves were dying. And where there had been freshwater marshes, the sea had come in and the salt had brought death.

But with the death and the dying grasses there was a rebirth in the process. The old had been killed out and the new growths were springing up with vigor. In a year or two there would be no sign of the hurricane. All the arrowhead weeds and river swamp vegetation far up that river that was killed would once again be green and full of life. There were whole sections of the marshlands that were filled in where sand dunes had washed back and built up the land. The life-building process goes on and on, but nothing lasts forever, not even our pollution and our alteration of nature and her environment.

As the land began to rebound, people did also. There was talk of rebuilding the houses on Alligator Point that had been washed into the sea, this time on stilts, this time behind the dunes instead of in front of them. The sea had taught only respect. She had chosen not to teach fear and terror this time. The Office of Emergency

Planning had set up an office in town. The government was going to spread loans out to rebuild and restore. My gloom and depression of the last few days had turned to hope. *Penaeus* had managed to ride out the storm without a scratch, and before long we were back collecting and shipping out orders. Already we were making plans for a new dock and getting estimates. Now I could rebuild like I never could have afforded to do in the first place. Now I could have a super dock, with more floating boat stalls and better culture racks and facilities.

An engineer from the Small Business Administration came to visit. We showed our twisted wreckage and talked about what we had lost. "You talk about that dock like it was a person that was killed or something," he said, shaking his head. "It was only a dock, for God's sake."

But how do you make someone understand that all those barnacles and encrusting creatures were valuable and important, and that the dock was the greatest place in the world? How do you tell him that people no longer have a place to fish, and that we have no place to tie up our boats, and that something cultural and great has been lost? The engineer didn't have much time. He went from house to house, from structure to structure, writing down general estimates. He didn't have a whole lot to say. He listened to people with suspicion. He wanted to be fair, but he had to protect the taxpayers' money. He understood the crab house that had a wall knocked down by the waves and had their cans and packaging materials scattered to the wind and saturated and ruined by salt water.

He inspected Mr. Sebro's refrigerator and hot-water heater, which had been soaked with salt water, and listened to his claim that even though they were running, it wouldn't be long before they were ruined. He listened sympathetically as Mrs. Johnson showed him her water-soaked rugs, and he said nothing when I showed him the twisted rubble of our sea water line that kept our tanks nourished with specimens. I was a special case, he said. He had no experience to deal with me. He had never seen anything like my operation.

The more I spoke to him, the more depressed I became. I was

especially depressed when I came back from the emergency office with a huge stack of forms that had to be filled out, and I thought of having to remortgage my already mortgaged soul, even at low interest, for the next thirty years. There was even a considerable question as to whether or not I would be eligible for a loan. They loaned money to viable businesses and Gulf Specimen Company couldn't exactly be considered viable. It had never really made a profit in all its years of operation.

I worked day and night with my accountant, filling out the forms, trying to understand the questions. I submitted the forms when I had done all I could with them, and they came back from the Small Business Administration in Jacksonville with hordes of questions. What kind of business were we? What did we do? What were we really up to? There was something suspicious about it. If we were a seafood house, a service station, a plumbing shop with a traditional type of operation, the Small Business Administration could do something about it. Gulf Specimen Company required a specialist before they would release the thousands of dollars necessary to put the pieces back together again.

An old man named Mr. Smith arrived unannounced in the laboratory one day. He was the senior loan officer and it was on his say-so that our loan would fly or not. He walked around the lab, looked in the offices, peered into the tanks, and walked into the central packing room where our two pet sea turtles swam over to him with their mouths open, hoping he would feed them. Always they begged for food, and sometimes it worked.

Once again I explained our operation. I showed him our catalogs. Mary Ellen handed him a purchase order from the University of Illinois and then one from Harvard Medical College. Anne showed him the invoices and explained about the value of horseshoe crabs and how they were someday going to be a great tool in medical science, and Leon told him how every year hundreds of schoolchildren came by to look at the facilities. He said little. He just listened.

He spread out my financial statement and frowned and shook his head. He ran his fingers down the column. "Mr. Rudloe," the old man finally said, shaking his head, "I've been with the SBA for nearly forty years now. I have seen every kind of business that you can imagine. I have seen sausage factories and plants that make the Styrofoam boxes that you ship your specimens in. I have seen thousands of seafood operations. I don't know what to say to you, sir. Never in all my years of service to the government have I seen a business like Gulf Specimen Company. You obviously provided a much needed service, but a business has to make a profit. If I were to go strictly by the rules I would have to turn down this loan."

He paused, and I could see that he was shaken, that somewhere deep down in his bureaucratic mind that functioned on form 403A in triplicate, he saw that there was no order or reason for my strange business that sold the wonderful life of the sea. How could a business with a balance sheet like Gulf Specimen's exist? Why hadn't it become extinct long ago? Yet exist it did.

"Mr. Rudloe . . . " His voice wavered. "I'm going to authorize your loan, in the full amount. You will be able to rebuild your facilities and replace your dock. I'm not sure what you have claimed here is what it should be, but I am certainly not the one to say whether it is right or wrong."

We shook hands and he started out the door, and then he turned to me. "By the way, Mr. Rudloe, don't you think you ought to feed your turtles out there? They look terribly hungry."

# Index

**JACK RUDLOE** was born in New York City. In 1957 he moved to Panacea, Florida, where he became an environmental activist and founded Gulf Specimen, a marine biological-supply company that collects and ships sea creatures, both humble and exotic, to laboratories and aquariums throughout the country. He has participated in a number of collecting expeditions for the New York Aquarium, the Smithsonian and other institutions, and was a member of the International Indian Ocean Expedition to Madagascar. His articles have appeared in many notable publications, including *Audubon, National Geographic* and *Natural History.* Among his books are *The Sea Brings Forth, Time of the Turtle* and *The Wilderness Coast.* He still lives in Panacea, with his wife, Anne, and their two children.

**WALTER INGLIS ANDERSON** is a Mississippi artist who, since his death in 1965, has been hailed as a genius—yet he died virtually unknown. A recluse and a mystic who chose to live in the world of nature rather than of man, he fought to convey the living essence of the natural world around him, to bring "nature and art into one thing." Not until his death did his family discover his voluminous journals and the multitude of watercolors, pen-and-ink drawings, block prints, wood carvings and sculptures he had left behind. His reputation today is based primarily on the tens of thousands of watercolors and the extraordinary murals in his home town of Ocean Springs, Mississippi.

## The Living Dock

Designed by Bryan Dahlberg
Composed in ITC Galliard by LaserWriting, Inc.,
Englewood, Colorado, on a Linotronic 300
Color printing by Strine Printing Company, York, Pennsylvania
Color separations by Spectrum, Inc., Golden, Colorado
Printed and bound by Arcata Graphics, Fairfield, Pennsylvania
Printed on Sebago Antique, acid-free paper